International Political Economy Series

Series Editor: **Timothy M. Shaw**, Visiting Professor, University of Massachusetts, Boston, USA, and Emeritus Professor, University of London, UK

The global political economy is in flux as a series of cumulative crises impacts its organisation and governance. The IPE series has tracked its development in both analysis and structure over the last three decades. It has always had a concentration on the Global South. Now the South increasingly challenges the North as the centre of development, also reflected in a growing number of submissions and publications on indebted eurozone economies in Southern Europe.

An indispensable resource for scholars and researchers, the series examines a variety of capitalisms and connections by focusing on emerging economies, companies and sectors, debates and policies. It informs diverse policy communities as the established trans-Atlantic North declines and "the rest", especially the BRICS, rise.

Titles include:

Andrei V. Belyi
TRANSNATIONAL GAS MARKETS AND EURO-RUSSIAN ENERGY RELATIONS

Silvia Pepino
SOVEREIGN RISK AND FINANCIAL CRISIS
The International Political Economy of the Eurozone

Ryan David Kiggins (*editor*)
THE POLITICAL ECONOMY OF RARE EARTH ELEMENTS
Rising Powers and Technological Change

Seán Ó Riain, Felix Behling, Rossella Ciccia and Eoin Flaherty (*editors*)
THE CHANGING WORLDS AND WORKPLACES OF CAPITALISM

Alexander Korolev and Jing Huang
INTERNATIONAL COOPERATION IN THE DEVELOPMENT OF RUSSIA'S FAR EAST AND SIBERIA

Roman Goldbach
GLOBAL GOVERNANCE AND REGULATORY FAILURE
The Political Economy of Banking

Kate Ervine and Gavin Fridell (*editors*)
BEYOND FREE TRADE
Alternative Approaches to Trade, Politics and Power

Ray Kiely
THE BRICS, US "DECLINE" AND GLOBAL TRANSFORMATIONS

Robin Bush, Philip Fountain and Mark Feener (*editors*)
RELIGION AND THE POLITICS OF DEVELOPMENT
Critical Perspectives on Asia

Markus Fraundorfer
BRAZIL'S EMERGING ROLE IN GLOBAL SECTORAL GOVERNANCE
Health, Food Security and Bioenergy

Katherine Hirschfeld
GANGSTER STATES
Organized Crime, Kleptocracy and Political Collapse

Matthew Webb and Albert Wijeweera (*editors*)
THE POLITICAL ECONOMY OF CONFLICT IN SOUTH ASIA

Matthias Ebenau, Ian Bruff and Christian May (*editors*)
STATES AND MARKETS IN HYDROCARBON SECTORS
Critical and Global Perspectives

Jeffrey Dayton-Johnson
LATIN AMERICA'S EMERGING MIDDLE CLASSES
Economic Perspectives

Andrei Belyi and Kim Talus
STATES AND MARKETS IN HYDROCARBON SECTORS

Dries Lesage and Thijs Van de Graaf
RISING POWERS AND MULTILATERAL INSTITUTIONS

Leslie Elliott Armijo and Saori N. Katada (*editors*)
THE FINANCIAL STATECRAFT OF EMERGING POWERS
Shield and Sword in Asia and Latin America

Md Mizanur Rahman, Tan Tai Yong, Ahsan Ullah (*editors*)
MIGRANT REMITTANCES IN SOUTH ASIA
Social, Economic and Political Implications

Bartholomew Paudyn
CREDIT RATINGS AND SOVEREIGN DEBT
The Political Economy of Creditworthiness through Risk and Uncertainty

Lourdes Casanova and Julian Kassum
THE POLITICAL ECONOMY OF AN EMERGING GLOBAL POWER
In Search of the Brazil Dream

Toni Haastrup, and Yong-Soo Eun (*editors*)
REGIONALISING GLOBAL CRISES
The Financial Crisis and New Frontiers in Regional Governance

Kobena T. Hanson, Cristina D'Alessandro and Francis Owusu (*editors*)
MANAGING AFRICA'S NATURAL RESOURCES
Capacities for Development

International Political Economy Series
Series Standing Order ISBN 978–0–333–71708–0 hardcover
ISBN 978–0–333–71110–1 paperback

You can receive future titles in this series as they are published by placing a standing order. Please contact your bookseller or, in case of difficulty, write to us at the address below with your name and address, the title of the series and one of the ISBNs quoted above.

Customer Services Department, Macmillan Distribution Ltd, Houndmills, Basingstoke, Hampshire RG21 6XS, England

Alliance Capitalism, Innovation and the Chinese State

The Global Wireless Sector

Victoria Higgins
Monash University, Victoria, Australia

First published 2015 by
PALGRAVE MACMILLAN

Palgrave Macmillan in the UK is an imprint of Macmillan Publishers Limited, registered in England, company number 785998, of Houndmills, Basingstoke, Hampshire RG21 6XS.

Palgrave Macmillan in the US is a division of St Martin's Press LLC, 175 Fifth Avenue, New York, NY 10010.

Palgrave Macmillan is the global academic imprint of the above companies and has companies and representatives throughout the world.

Palgrave® and Macmillan® are registered trademarks in the United States, the United Kingdom, Europe and other countries.

ISBN: 978–1–137–52964–0

This book is printed on paper suitable for recycling and made from fully managed and sustained forest sources. Logging, pulping and manufacturing processes are expected to conform to the environmental regulations of the country of origin.

A catalogue record for this book is available from the British Library.

Library of Congress Cataloging-in-Publication Data

Higgins, Victoria, 1970–
    Alliance capitalism, innovation and the Chinese state: the global wireless sector / Victoria Higgins.
        pages cm.—(International political economy series)
    ISBN 978–1–137–52964–0 (hardback)
    1. Technological innovations – Economic aspects – China. 2. Information technology – Economic aspects – China. 3. High technology industries – China. 4. Telecommunication – China. 5. Public-private sector cooperation--China. 6. Strategic alliances (Business) – China. 7. Technology and state – China. I. Title.
HC430.T4H55 2015
384.5'30951—dc23   2015015594

Contents

Acknowledgements

This book would not have been possible without the support I have received from many people. I would like to thank Professors Stephen Bell and David Jones for their guidance, support and breadth of knowledge. Professor Bell read and commented on the entire typescript, for which I am truly grateful.

I would like to convey my thanks to the University of Queensland, specifically the School of International Relations and Political Science, which has provided the funding and facilities required to complete my research. I also would like to thank the committee of the Chinese Goes Global conference held at Harvard University in 2011 for providing the opportunity to present and refine my ideas. I thank Hu Biliang, Dean of Beijing Normal University, for his invitation to present and further refine my work as a guest lecturer at Beijing Normal University in 2012. Thanks also to my interviewees in China for their time and pertinent insights into China's innovation process. Finally, I wish to express my love and gratitude to my beloved family, for their understanding and endless support, throughout the duration of the writing process.

List of Abbreviations

APS	application providers
BPO	business process outsourcing
BWA	broadband wireless access
C & D	concept and develop
CASS	Chinese Academy of Social Science
CCP	Communist Party of China
CCSA	China Communications Standardization Association
CDMA	Code Division Multiple Access
CGINs	collaborative global innovation networks
CIC	China Investment Corporation
CIT	Corporate Income Tax
CPs	content providers
DCI	Digital Cinema Initiative
DJIA	Dow Jones Industrial Average
EDA	electronic design automation
EMS	electronic manufacturing services
EPs	equipment providers
ETSI	European Telecommunications Standards Institute
ETDZ	economic and technological development zones
FDD	frequency-division duplexing
FDI	foreign direct investment
FFEs	foreign-funded enterprises
FIE	foreign-invested enterprise sector
FRAND	fair, reasonable and non-discriminatory
FYP	Five-Year Plan
GDP	gross domestic profit
GFC	global financial crisis
GIPN	global innovation production networks
GPN	global production networks
GSM	Global System for Mobile Communications
GTI	Global TD-LTE Initiative
HDTV	high-definition television sets
HSPA	High Speed Packet Access
3G	Third Generation Mobile Telecommunications Technology Standards
3GPP	Third Generation Partnership Project

4G	Fourth Generation Mobile Telecommunications Technology Standards
IEEE	Institute of Electrical and Electronics Engineers
IMF	International Monetary Fund
IPO	initial public offerings
IPR	Intellectual Property Rights
ISO	International Organisation for Standardisation
ITO	information technology outsourcing
ITU	International Telecommunication Union
JVs	joint ventures
KPO	knowledge process outsourcing
LSTI	LTE/SAI Trial Initiative
LTE	Long-Term Evolution
MII	Ministry of Industry and Information
MOF	Ministry of Finance
MOFCOM	Ministry of Commerce
MOFTEC	Ministry of Foreign Trade and Economic Cooperation
MOST	Ministry of Science and Technology
MNE	multinational enterprise
MPT	Ministry of Posts & Telecommunications
MPS	megabits per second
NGMN	Next Generation Mobile Networks
NDRC	National Development and Reform Commission
NPC	National People's Congress
NSN	Nokia Siemens
NVCA	National Venture Capital Association
OECD	Organisation for Economic Co-operation and Development
OEM	original equipment manufacturer
OFDI	outward foreign direct investment
OMA	Open Mobile Alliance
OPA	open patent alliance
PC	Peoples Congress
PCT	Patent Cooperation Treaty
PE	private equity
PHS	Personal Handy-phone System
P & G	Procter and Gamble
PRC	People's Republic of China
PPF	partially privatised firm
PTTS	postal, telegraph, and telephone service

SAAC	State-owned Assets Supervision and Administration Commission
SEZs	special economic zones
SIPO	State Intellectual Property Office
SKT	SK Telecom
SOE	state-owned enterprise
SOHC	state-owned holding company
SPs	service providers
STI	science, technology and innovation
SWF	Sovereign Wealth Fund
TBT	Technical Barriers to Trade
TDIA	TD Industry Alliance
TD-LTE	Time-Division Long-Term Evolution
TD-SCDMA	Time Division Synchronous Code Division Multiple Access
TM	TeleManagement Forum
TNC	transnational corporations
TSMC	Taiwan Semiconductor Manufacturing Company
UMTS	Universal Mobile Telecommunications System
VC	venture capital
WAPI	Wireless LAN Authentication and Privacy Infrastructure
WCDMA	Wideband Code Division Multiple Access
WFOE	wholly foreign-owned enterprises
WiMAX	Worldwide Interoperability for Microwave Access
WIP	Wireless Industry Partnership
WIPO	World Intellectual Property Organization
WTO	World Trade Organization
WWRF	Wireless World Research Forum

1
China's New Alliance Capitalism and the Case of the Wireless Communication Sector

Hailed as an economic miracle, the People's Republic of China (PRC) has become a key example of state-led, market-oriented global economic reform (Enright, Scott and Chang, 2005:1). The virtues and shortcomings of China's economic rise are touted by economists and policymakers worldwide. For some, the economic re-emergence of the Chinese nation is one of the most significant events to occur in modern history (OECD, 2006). For others: "China is an object lesson in the threat that central-ised, authoritarian states pose to revolutionary technological develop-ment" (Freeland, 2010).

It is certainly undeniable that since the inauguration of economic reforms in 1978, with Deng Xiaoping's "Open Door Policy", China's economic performance has been unprecedented in "speed, scale and scope" (Pei, 2006:143). This policy has been a pragmatic economic reform program, revolving around the need to generate capital resources to finance the modernisation of the Chinese economy. It encouraged the formulation of rural enterprises and private businesses, liberalised foreign trade and investment, relaxed state controls over some prices, and invested in industrial production and the education of its workforce (Pei, 2006:2).

Three principal factors have influenced the process of post-Mao economic reform in China. These are its focus on path dependency, developmental style state-led industrialisation, and the reinvention of the Chinese Communist Party (CCP) and its monopolistic hold on power. Piecemeal social engineering in the formative stages of market transition led the central state inexorably to oversee institutional changes to establish a modern legal-rational bureaucracy. Although the state remains structurally vulnerable to rent seeking, it gained the organisational capacity to institute and enforce rules critical to the emergence of a hybrid market economy (Enright, Scott and Chang, 2005:1).

The growth of the Chinese economy since the onset of reform has averaged 9.1% per annum income and GDP has quadrupled in the last 15 years (Enright, Scott and Chang, 2005:1). China's economic development has lifted more people out of poverty than in any other nation in a comparable time frame (Enright, Scott and Chang, 2005:1). This impressive growth is a direct product of a flourishing labour-intensive domestic manufacturing industry, which has made China "a world factory of low cost goods" (Bell and Feng, 2006:53).

However, it is not the objective of Chinese state leaders to build just another workshop to fulfil the consumption aspirations of the developed world. To the contrary, China's leaders are intent on flying aeroplanes made in Shanghai, using computers designed and built in Beijing and driving automobiles that have been manufactured in Guangzhou (Thun, 2006:3). Indeed, an elite consensus, it seems, has crystallised around the full-scale mobilisation of resources for technological catch-up and for architectural platform leadership in the next generation of high technologies. Hence, "Innovation with Chinese Characteristics" has become a new mantra that serves to frame the science and technology policymaking directives of government authorities in the twenty-first century.

It is important to note here that the exact tenants of this "grand innovation" and "high-technological development strategy", specifically, its "go-it-alone" technological development strands, have been the subject of significant revision since its initiation in 2006 as the government of China adapts to both socio-technological systems change and geopolitical bargaining processes. For instance, early versions of the Chinese government's innovation development plan were issued in 2006 and primarily operated from the assertion that the ability of the Chinese nation to transcend its position as a low-cost producer of modular, undifferentiated manufactured goods in the global economy and become a high-tech, innovative knowledge economy required the construction of a sophisticated indigenous technology base, capable of challenging the economic and technological supremacy of foreign multinational companies (MNCs). This policy direction can be seen exemplified by the launch of the January 2006 "National Medium and Long Term Program (MLP) for Scientific and Technological Development" (S&T) (2006–20). The MLP (2006–20) called for China to become an "innovation-oriented society" by the year 2020 and a world leader in S&T by 2050. It committed China to developing capabilities for "indigenous innovation" (*zizhu chuangxin*) and to leapfrog into leading positions in new science-based industries by the end of the plan period.

However, whilst it is unarguable that the plan contained a distinctive techno-nationalistic thrust, it was also defined by concern about the ability of the Chinese nation to accomplish its key technological development goals on its own. For example, the plan acknowledged that the Chinese S&T system was defined by a number of shortcomings. Specifically, it was asserted that the innovative capability of Chinese enterprises was "weak", the S&T sector too compartmentalised, and management of the S&T system was "terribly uncoordinated". It also conjectured that the system failed to award high achievers or encourage innovation more generally. Hence, the plan concluded that China's indigenous innovation and technological development process would need to be supplemented with international cooperation (Kennedy, 2013). Indeed, as this book will highlight, due to domestic and international political lobbying, the complexity and cost of technological innovation and the increasingly networked and relational nature of contemporary global innovation and production networks, technological development programs in the global wireless communication sector launched by the Chinese government that were overtly "go-it-alone" and "techno-nationalistic" have essentially failed.

Indeed, it is becoming increasingly clear that in this next round of technological development the existence of critical ecosystem dependencies in many high-technology sectors is facilitating a shift towards a more collaborative, alliance-based socio-technological development paradigm. As a direct consequence, the strategic behaviour of high-technology actors, including firms and governmental policy actors, is currently being reframed from "go-it-alone, techno-nationalistic" policy agendas to more "open globalised co-development models". This is because in the contemporary socio-technological environment "go-it-alone" and "indigenous development agendas" with a techno-nationalistic orientation are unable to develop the necessary relational ties and ecosystem embeddedness necessary for the development and commercialisation of high-technology assets.

Certainly, as this book will highlight, the rapid pace of technological change and the increasing organisational and relational complexity of doing business in these high-technology sectors make it impossible for a single company – or even a single nation – to compete without technological, financial and organisational collaboration. What I want to emphasise here is that the ways in which contemporary high-technology sectors are both nested within and are highly dependent on complex crisscrossing infrastructures, combined with the rapidly accelerating pace of technological change at multiple levels, indicates that "technology" itself is a "major agent" of both "system change" and "structuring" in the contemporary socio-technical system. Specifically, the concepts of "convergence" and "interoperability" are "key structural change drivers" that facilitate the development of these critical ecosystem dependencies and play a fundamental role in shaping the behaviour of system actors and hence need to be accounted for in theoretical and empirical modelling. This is because in order to participate in the global networks and sectoral systems that comprise the contemporary socio-technical system, actors need to develop technological innovations that possess the ability to converge or interact with other firms' products. Furthermore, the ability to ensure "convergence" and "interoperability" is highly dependent on the capacity of firms to access globally dispersed scientific and technical knowledge sources and to generate long-term networked relational ties with other system actors.

However, this does not mean that actors are unable to shape their environment in order to develop or appropriate high-technology assets. To the contrary, the need for technical infrastructure and products that are designed to ensure system "convergence" and "interoperability" means that innovative technological development is defined not just by spontaneous evolution and organisation but also by directed and coordinated organisation by goal-oriented system actors. That is, technological sectors, their supportive structures, technology-specific institutions and corresponding ecosystems are primarily the result of deliberate actions and policy choices made by innovating system actors. Moreover, because the contemporary socio-technological environment is defined by the need for "interoperability" and "ecosystem embeddedness", the sets of policy choices open to innovating system actors are framed by the need to ensure the development of both relational assets and collaborative technological development at multiple levels and with multiple actors.

As a consequence, the outline of a policy shift towards a more collaborative, innovation-based, high-technology development plan is emerging in China. Whilst the exact parameters of this policy framework are in its early stages, its overarching thrust is structured around the need to develop the necessary institutional, organisational, relational and research and development (R&D) capabilities in order to facilitate the access of Chinese firms to the contemporary global economies horizontally networked, geographically dispersed and increasingly partnered innovation networks and processes.

Key objectives

The primary goal of this book is to highlight how technological change and global systems integration via the development and proliferation of scientific and organisational innovation networks, the occurrence of critical ecosystem dependencies and the emergence of new forms of collaborative architecture are generating a fundamental shift towards a more alliance-based model of capitalist organisation.

In this book the strategic response of the Chinese government and its key domestic firms to the emergence of this more collaborative, alliance-based, socio-technological economic system will be examined. The empirical focus here will be on the 3G and 4G wireless technology sector. It is important to note here that whilst it can be argued that this more alliance-based form of capitalism is also evident in a number of other sectors such as aerospace, biotechnology, nanotechnology, the automobile sector, and the information and computer technology sector (ICT), the focus of this book is confined to the global wireless communication sector. Specifically, this book will highlight how, in order to participate in the development of the 3G and 4G wireless ecosystem and gain core technology and market share, the Chinese government has undergone a shift in its political agenda, technological focus and R&D expenditure allocations. I will highlight how the government of China has changed its technological development approach and moved away from its earlier techno-nationalistic, indigenous, innovation development agenda in the sector and sought to forge and reinforce far-reaching global alliances with foreign and domestic multinationals.

Beyond the earlier foreign direct investment (FDI) and techno-nationalist strategies, this new development strategy can be conceptualised as the "third major attempt" by the Chinese government to employ an effective technology development strategy in China. The new strategy's key strength is the way in which it endeavours to reconcile the prominent Chinese New Left faction's techno-nationalistic ambitions with the liberal desire to utilise the contemporary globalisation process as a mechanism to facilitate growth and technological innovation-oriented upscaling. It is essentially a "third way" whereby the state's strategic control over the globalisation process is modified in a way that utilises socio-technological systems change to ensure that the interests of global firms are actually integrated and embedded into the Chinese high-technology developmental strategy. In this book, I will term this third strategic attempt to achieve high-technology-based development "globalised adaptive ecology". By this, I mean that Chinese policymakers have adopted a globalised adaptive approach to high-technology upscaling that is both responsive to the fact that contemporary technological knowledge is located in multiple knowledge zones and technical ecosystems, and the fact that these are highly changeable and can be shaped and structured by goal-directed system actors. By seeking to exploit the way in which globalised technological ecosystems adapt and change over time, with their continuous shedding and need for new participants, Chinese policymakers for the first time have been able to create a platform for the nation to achieve high-technology upscaling and innovative product development on a global scale.

In this new model of Chinese technological development, a dynamic model of market competition and cooperation directs firms towards more flexible quasi-market cooperation models that are designed to enable them to experiment with new ideas, pre-competitive market construction and product markets in a lower risk regime. The parameters of this new technological policy program are its emphasis on a hybrid mixture of ownership and corporate governance patterns as well as a set of aggressive policies designed to foster alliances and co-development platforms with global leaders in industry and R&D (Ernst and Naughton, 2008:40).

The overarching result has been the increasing internationalisation of the Chinese state and its domestic firms into trans-territorial technological ecosystems and knowledge zones. The success of this strategy has been highly dependent on the ability of state leaders to move beyond contentious bargaining frameworks where the overarching goal was to secure respective sectoral and industry concessions for Chinese domestic firms and its replacement with a more collaboratively oriented model of co-development that focuses on achieving win-win scenarios for both foreign MNCs operating in China and domestic firms wishing to embed themselves in global production and innovation networks.

From a policy perspective, this third strategic developmental attempt in the sector has been highly successful. Key collaborative capability has been built up in the sector nationally and globally by both Chinese state firms and domestic firms operating in the sector and new policy tools for the interlinking of domestic and foreign actors together in a way that facilitates the co-production, and appropriation of technological assets are beginning to evolve. The overarching result I will argue is a fundamental shift in Chinese governmental policy directives from one that was primarily focused on commanding specific outcomes to an approach that is designed to engage in the creation and maintenance of new markets, and a move away from top-down policy development to one that is focused around steering and negotiating alliances with partners in both the domestic private sector and the international sphere.

The emergence of a more collaborative, alliance-based global economy and its impact on Chinese high-technology developmental policy has yet to be addressed in a comprehensive fashion by the international development and globalisation literature. This book will attempt to address this gap and highlight how this shift towards a more collaborative socio-technological environment has fundamental implications, not just for the Chinese state and its domestic firms but also for the global high-technology system itself, its system of intellectual property rights, its relational networks and modes of value appropriation.

The remainder of this introductory chapter is comprised of nine sections. In the first and second sections, I will provide a brief overview of the idea of critical ecosystem dependencies and how they relate to the emergence of alliance capitalism as an emerging development strategy. In the third section, I will introduce the idea of complex adaptive systems and highlight how recent socio-technological change impacts upon the ability of systems actors to shape their technological developmental processes. I will highlight how key system actors are addressing the existence of these emergent critical ecosystem dependencies by adopting collaborative alliance structures, and I will argue that a defining feature of the emerging socio-technological system is the purposeful construction and coordination of high-technology ecosystems via strategic agency. I will introduce the concepts of ecosystem shaping, co-development and system embeddedness as new strategic models of behaviour that are being increasingly employed by system actors to facilitate successful, globalised, high-technological development at both the national and global level. In section four, I will explore the way in which these socio-technological system changes impact on the state and its role in the contemporary technological development processes. In sections five and six, I will introduce the global network state (GNS) developmental model as a theoretical framework from which to examine China's state capacity at the sectoral level and highlight how a key role of the Chinese government as a developmental state in the contemporary wireless technology sector is not just confined to fixing market failures, but also involves the active creation of markets for new technologies. In sections seven and eight, I will outline the analytical framework's key concepts and methodology, and in the final section I will provide a brief chapter outline of the book.

Critical ecosystem dependencies and the global knowledge economy

The concept of the knowledge-based economy began to emerge in the early 1970s, and has since evolved into both an explanatory and norma-tive framework for examining the emergence of innovative technolo-gies and intangible assets (Schilirò, 2010). Intangible assets are assets that do not have a physical embodiment. Termed "intellectual assets" by the Organisation for Economic Co-operation and Development (OECD, 2011), intangible assets have also been referred to as knowledge assets or intellectual capital. Much of the focus on intangibles has been on R&D, key personnel and software. But the range of intangible assets is significantly broader. For instance, one classification groups intangibles into three primary types: computerised information (such as software and databases); innovative property (such as scientific and non-scien-tific R&D, copyrights, designs, trademarks); and economic competencies (including brand equity, firm-specific human capital, networks joining people and institutions, organisational know-how, adaptability, organi-sation capital that increases enterprise efficiency, and aspects of adver-tising and marketing) (OECD, 2011).

It is pertinent to note that the proportion of economic value that was attributable to the innovative capacity of intangible assets has grown significantly since the 1980s. For example, in 2005, intangible assets represented 80% of market value on the Dow Jones Industrial Average (DIJA). In stark contrast, in 1980, the DIJA reflected market values due to intangible assets at zero. Furthermore, it is interesting to note that the value of a number of leading multinational companies, such as Microsoft and Apple, is now almost entirely accounted for by their intangible assets alone (Kuznar, 2012).

The production of knowledge and intangible assets in the contem-porary global economy primarily occurs in innovation ecosystems and collaboratively managed networks where ideas and products are co-cre-ated and commercialised. Collaborative global innovation networks (CGINs) can be defined as: "A globally organised web of complex inter-actions between firms and non-firm organisations engaged in knowl-edge production related to and resulting in innovation" (Chaminade, 2009 cited in Barnard and Chaminade, 2012). Fragmentary evidence suggests most innovation networks are organised in complex techno-logical sectors. Moreover, it can be argued, that these networks and tech-nologies are defined by the way they co-evolve. In other words, changes in networks may lead to changes in the innovation of technologies;

and those technological changes may serve to modify the network itself (Rycroft, 2003).

Furthermore, it is important to note here that CGINs are not just defined by their spatially diverse and global nature, or by their political or geographical territorial jurisdiction. Instead they are represented by sites of codified knowledge, expertise and collaborative organisation. The jurisdictional fields of technology, knowledge, productiveness and innovation that regulation attempts to define can usefully be conceptualised as "zones" (Barry, 2006). The definition of a technological zone has some flexibility, asserts Barry. For example, it may or may not be commensurate with a political territory, be it a nation state or other form or level of jurisdiction. Zones, it has been argued, make association between participants possible but also create new distinctions and separations. As Barry articulates zones are "spaces of circulation in which technologies take more or less standardised forms and in which intellectual property implies new 'objects of technical practice'". Political actors such as states, it is conjectured, have a key role in drawing and legitimating these boundaries and entry criteria (Barry, 2006).

This strategic shift in the behaviour of the firm, from an overt focus on the development and protection of proprietorial ideas and standards to a focus on the need to embed themselves in globalised innovation ecologies, has important implications for how firms and states can develop and govern the technological development process in the future.

I will argue that a key change that needs to be accounted for in theoretical and empirical modelling is the emergence of sets of critical ecosystem dependencies that policy and market actors need to address when developing policy directives and development plans. The primary point I want to emphasise here is that in many high-technology sectors the development of high-technology assets and their successful commercialisation is highly reliant on the existence of an ecosystem of partners that are both able and willing to participate in the development of complementary products and service solutions. For example, in order for Airbus to bring its super jumbo passenger aircraft, the A380, to market, an extensive ecosystem of interdependent actors and linkages needed to be constructed. This is because Airbus is highly dependent on a host of suppliers for sub-assemblies and components (navigation systems, engines, etc.). Furthermore, some of these suppliers were themselves confronted with significant innovation challenges as they attempted to develop the components required to build the airbus. Moreover, it is imperative that Airbus integrates these components effectively into the core ecosystem and provides the necessary relational and technological support to its suppliers in order that they can adapt to the innovation challenges they face. Hence, it is not just suppliers that need to be considered when constructing a functioning ecosystem. It is also important that complementary service providers are also integrated into the overarching ecosystem (e.g., airports that need to invest in new infrastructure to accommodate the oversized planes, regulators who oversee safety specification and compliance issues, etc.) (Bauer, Lang and Schneider, 2012).

Conversely, it can be argued that technological components and assets are always embedded and functionally integrated in a larger socio-technological system. Moreover, technological development in the contemporary socio-technological systems is comprised of an array of nested systems, each performing independent functions that are integrated into a complex, interdependent, technological ecosystem (Hobday, 1998). This nested systems perspective acknowledges that the degree of advancement of a technology is not only related to the sophistication of its individual components but also to the complexity of its relational and functional integration. A technological system thus is not only defined by the construction of its various parts but also through the coordination and integration of a heterogeneous complex of multiple components and subsystems into an overall functioning technological ecosystem. Hence, technological advancement does not only relate to technological artefacts in the narrow sense, but also includes arrays and networks of social technologies, such as organisational patterns, logistical systems, and complex forms of social coordination. Since the 1980s, such integrated technological systems have also been conceptualised as "large technical systems" based on multiple technical and social components (Hobday, 1998).

As a direct consequence, it is becoming increasingly clear that understanding firm performance and technological governance in such "innovation ecosystems" requires a change in the way in which the strategy and the innovation literatures have traditionally linked industry dynamics to firm performance (Adner and Kapoor, 2006). Primarily, it requires an approach that is designed to address not only the innovation challenges that are faced by the focal firm (Cooper and Schendel, 1976; Tushman and Anderson, 1986; Henderson and Clark, 1990; Christensen, 1997 cited in Adner and Kapoor, 2006), but one that also considers the nature of the innovation challenges confronted by the external ecosystem partners. "Moreover, it requires an approach that extends beyond the focus on how different actors will bargain over value capture (Porter, 1980; Teece, 1986; Brandenburger and Stuart, 1996; Brandenburger and Nalebuff, 1997 cited in Adner and Kapoor, 2006) to include an explicit consideration of the innovation challenges that different actors will need to overcome in order for value to be created in the first place" (Adner and Kapoor, 2006).

Alliance capitalism

In this book I use the concept of alliance capitalism primarily to refer to the strategic policy challenges emerging in the global economy that require government actors and firms to develop embedded relational ties and collaborative R&D activity with other firms and economic and technological actors in order to engage in innovative upscaling and product development.

These alliances, it will be argued, are not just designed to appropriate complementary assets and resources as earlier alliance forms did, but are designed to facilitate the development of other key system actors in order to ensure that critical infrastructure, regulatory standards and complementary technologies are developed before a technology goes to market. These alliances can be termed innovation ecosystem-building alliances. The primary goal of such alliances is to anticipate future market and ecosystem requirements and to use this information to build a critical network of interdependent alliance partners that are focused on achieving technological "convergence" and "interoperability" across the ecosystem platform.

These alliances are highly interdependent and co-evolve as a complex adaptive system. This is because both the alliances and the socio-technological system are engaged in a process of self-organisation and evolution in a bid to adapt to and shape their changing environment. Strategic alliances can also be conceptualised as open systems (Katz and Khan, 1966). Open systems are defined by the way they are interconnected with other systems and exchange resources, and information with the external environment in which they are embedded (Katz and Khan, 1966). The overarching result is that traditional hierarchal capitalist forms of organisation and competitive market behaviour have been fundamentally altered as firms adapt to the emerging collaborative economy and the increasing need to engage in strategic ecosystem alliances in high-technology sectors. The adaption of the capitalist system itself can be seen manifest in the development of new, more open forms of intellectual property ownership and appropriation and the increasing propensity of firms to co-develop and co-own high-technology assets.

For developing nations such as China the occurrence of extraterritorial alliance structures has provided an important point of entry into global knowledge and production networks. It also expands the techno-logical governance arena in which it operates from one focused solely on achieving national development goals to one that is focused on achieving ecosystem development goals. From this frame of reference, effective technological governance in a contemporary context is inher-ently globalised and needs to take into account not just the interests of nation firms but also those that Chinese domestic firms are linked to by critical ecosystem dependencies.

Structure and agency in the global collaborative economy

The purposeful construction and coordination of high-technology ecosystems via strategic agency is a key feature of the contemporary socio-technological system. This is because in an emerging technological field, supportive structures and technology-specific institutions can neither be taken as given nor regarded as being external to technology develop-ment. Instead, they are often deliberately created and coordinated by innovating actors. Hence, "strategy" can itself shape "structure". This ability to "strategically shape" the socio-technical environment requires an awareness of deep structures, technological interoperability and the emergence and creation of new markets and organisation methods, and the ability to participate and coordinate activity in multiple technolog-ical arenas simultaneously.

The increasing need for technological ecosystem construction and maintenance, it can be argued, has created a fundamental need for antic-ipatory technological governance and coordination. By anticipatory technological governance I mean the development and execution by nation sates of a globalised technological development and integration strategy that attempts to construct future technological ecosystem plat-forms via the provision of basic research funding, long-term, strategic policy planning and collaborative alliance building. The overarching goal here is to build up a critical mass of trans-territorial technological system connections that the state can then utilise to facilitate and main-tain national innovative upscaling and appropriate globalised techno-logical assets and value.

The ecosystem metaphor provides a conceptual frame of reference that acknowledges the fact that technological ecosystems are in a constant state of flux as the ecosystem itself is continuously redefined. Hence, it is a metaphor that transcends simple unidirectional models of causality and development with the idea of complex interactional systems that are in a continuous process of adapting and growing. The metaphor can be used to describe existing conditions or those one might try to create; its users often aim to provoke new thinking about the conditions and requirements necessary to actively cultivate the development of an ecosystem to achieve a set of specific and desirable goals (Mars, Bronstein and Lusch, 2012).

For the purposes of this book, two main types of ecosystem shapers can be identified for analysis: focal organisers and network organisers. Focal organisers are focused on the provision of early funding and network development in order to facilitate what is termed "first to the world" basic research. The "World Wide Web" is an example of "first to the world" research (Atkinson, 2012).

In contrast, network organisers endeavour to exploit uncertainty in the ecosystem and build up a critical network of system actors in order to either disrupt earlier technological assets, form their own unique ecosystem spinoffs that serve to create new classes of technological assets and market structures, or build and add value to other system actors' established technological platforms. This kind of technological development strategy has been termed "innovation adaptation". It involves a process of taking a complex production system that is relatively well defined and building products and related processes and technological platforms (Atkinson, 2012). For example, the iPhone and Google both build upon earlier basic research findings to develop new products and technological assets. Furthermore, Apple is currently in the process of launching multiple new projects, enhancing the company's platform to enable more people to build business around Apple, for example, with iBeacons and a reconfiguration of the "iTV" ecosystem via:

> the release of a miniature device called "iRing" that will be placed on a user's finger and act as a navigation pointer for "iTV", enhancing the motion detection experience and negating some of the functionality found in a remote. Thirdly, "iTV" will come with a "mini iTV" screen that will seamlessly allow users to view content on this smaller, 9.7-inch screen, while also opening up use cases around home security, phone calls, video conferencing and other arena. (Shaughnessy, 2013)

The Chinese government and its private firms have yet to master the ability to produce "first to the world innovation" on a consistent scale (Atkinson, 2012), although key Chinese wireless technology firms such as Huawei and ZTE are succeeding in gaining indigenous technological knowledge and essential patents in the global wireless communication sector. Indeed, as highlighted in Chapter 5, Chinese domestic firms are highly adaptive fast followers that are achieving significant global market share in the 4G wireless communication sector as both ecosystem builders and network organisers via the development of new technological platforms, markets and collaborative alliance structures. From this perspective, achieving first-mover status is not the only way to move up the technology ladder and acquire market share. Instead, ecosystem shaping, co-development and system embeddedness are new strategic models of behaviour that can be utilised to facilitate successful globalised high-technological development at both the national and global level.

State capitalism and innovation

I have sought to highlight how high-technology development is becoming increasingly defined by the emergence of critical ecosystem dependencies and collaborative alliance structures. For the state, this changing socio-technological environment poses a number of important questions regarding its role in contemporary technological development processes and the trans-territorial appropriation and sharing of knowledge. For instance, should the state intervene to provide basic research funding in high-technology sectors with long-term time horizons which are unable to attract private capital due to risk aversion? Should it attempt to help facilitate the development of globalised relational networks and alliance structures or should it let them develop spontaneously? Can and should it play a role in sectoral coordination at a domestic and globalised level?

Obviously, the question of how much the state should intervene in the economy to facilitate high-technology economic development is a point of much theoretical contention. Free-market economists assert that government intervention should be strictly limited, as government intervention tends to cause an inefficient allocation of resources (Labonte, 2010). The close of the twentieth century was defined by the ideological precepts advanced by free-market theorists, specifically the theoretical premises embodied by the neoliberal free-market model. The neoliberal approach has strong theoretical premises: markets are efficient, the institutions needed to make markets work exist and are effective, and if there are deviations from optimality they cannot be remedied effectively by governments. The premises are a mixture of theoretical, empirical and political assumptions. Their theoretical core relies on, among other things, a restrictive view of the technological

basis of competitiveness. The empirical one relies on a particular interpretation of the experience of the most successful industrialising economies (Lall, 2004). Resource allocation is to be driven by the forces of free
markets, which operate to optimise a country's competitive advantage.
This, it is argued, will ensure dynamic advantage and yield the highest
rate of sustainable growth possible. In this approach, the only legitimate role for the state is to provide a stable macro economy with clear
rules of the game, open the economy fully to international product and
factor flows, give a lead role to private enterprise, and provide essential public goods such as basic human capital and infrastructure (Lall,
2004).

Neoliberal forces advocate the removal of capital controls and an end to
exchange rate manoeuvring in developing countries, as embodied in the
"Washington consensus" shared by the United States, the International
Monetary Fund (IMF) and the World Bank (Petersmann, 2005:128).
The prevailing perspective on financial liberalisation suggests a set of
standard procedures and rules for reform, including prudent regulation, transparent accounting and supervision, an orderly sequencing
of capital account liberalisation, and corporate restructuring. The legal
framework is that of a contextually negotiated contractual obligation
between juridical individuals in the marketplace. Private enterprise and
entrepreneurial activity are viewed as the keys to economic productivity
and wealth creation (Petersmann, 2005:128).

The idea advanced by the neoliberal free-market model that the forces
of supply and demand will spontaneously and freely produce prato-op-
timum outcomes and high growth rates without the need for organisational coordination or any legal and regulatory coercion has proved to
be inherently flawed. Certainly, it has become increasingly clear, that
neoliberalism's deregulatory policy prescriptions coupled with its key
theoretical precepts, specifically, the idea of perfect information, self-
regulatory markets and the existence of rational economic actors lead
to the 2008–09 global financial crisis (GFC). Furthermore, the high costs
of R&D, the increasingly short time frames for both product development and commercialisation, the occurrence of short product lifespans
and the significant decrease in venture fund capital since the financial
crisis means that the role of the state in high-technology development
is being reinvigorated as developing nations such as China engage in
a critique of leading Western international economic institutions and
free-market theoretical precepts. Indeed, it has been conjectured that a
shift towards the idea of state capitalism as a potent alternative model of
economic organisation has begun (Bremmer, 2009, 2012).

The idea that state capitalism is presenting a challenge to free-market principles is not a view without significant controversy. For example, Arbache (2013) argues that:

> state capitalist, Russia's economy is now being strangled by the state. Under Prime Minister (and formerly President) Vladimir Putin, the state reasserted its authority, regaining its dominance over key sectors of the economy, especially the crucial oil and gas industry. Putin also redistributed oil money by increasing government spending and the size of the civil service... State enterprises, favoured by over-bearing bureaucrats, are crowding out the private sector. World Bank surveys show Russia is becoming a harder and harder place to do business... Private capital is fleeing the country.... Even senior policy-makers within the Kremlin are doubting the future of Russia's state capitalist model. (Schuman, 2012)

In a similar vein, asserts Arbache:

> While the attractiveness of State capitalism is understandable within the context of economic crisis, its multiplication on a global scale has harmful implications. In fact, it seems to be highly unlikely that many countries will, simultaneously, benefit from State capitalism owing not only to the fallacy of composition, but also to the negative externalities brought on by them, which tend to upset the economic system, encourage trade and currency wars and raise political tensions between countries. (Arbache, 2013)

However, as highlighted in Chapter 3, the "successful" models of state capitalism being adopted by developing nations bear little resemblance to the command or nationalised economies of the past. In a contemporary context, state capitalism 2.0, as it has been termed (Musacchio, 2012), represents a new hybrid form of capitalism that is not defined by any single political or economic model, but by the way in which the state uses its power to foster economic development (Musacchio, 2012).

In state capitalism 2.0, the largest state-owned enterprises in the world are publicly traded and have large institutional investors monitoring their activities. Furthermore, big state-owned enterprises (SOEs) compete internationally, follow international reporting standards and have professional management, even in many public utilities, where social objectives commonly trump profitability. Recent reforms in companies like Indian Railways, Italy's Enel, or São Paulo's water and sanitation company, Sabesp, are some examples (Musacchio, 2012).

Indeed, it can be argued that classical interpretations of state capitalism structured around the idea that the state should not intervene in the market is not just clearly outdated but fundamentally flawed. For example, the World Wide Web was pioneered by the U.S. State Department, and the algorithm that provided the foundation for Google's success was funded by a public sector National Science Foundation grant. In a similar vein, molecular antibodies, which provided the foundation for biotechnology before venture capital moved into the sector, were discovered in public Medical Research Council (MRC) labs in the UK (Mazzucato, 2011). In her book, The Entrepreneurial State, Mariana Mazzucato asserts that lessons from these experiences are important precisely because:

> they force the debate to go beyond the role of the state in stimulating demand, or the role of the state in "picking winners" in industrial policy, where taxpayers money is potentially misdirected to badly managed firms in the name of progress, distorting incentives as it goes along. Instead it is a case for a targeted, proactive, entrepreneurial state, able to take risks, creating a highly networked system of actors harnessing the best of the private sector for the national good over a medium to long-term horizon. It is the state as catalyst, and lead investor, sparking the initial reaction in a network that will then cause knowledge to spread. The state as creator of the knowledge economy. (Mazzucato, 2011)

From this frame of reference, "innovation policy" is exactly where the Chinese government needs to play a fundamental role. Precisely because, without basic research funding, coordination and global alliance building and networking, high-technology innovation will often fail to occur at all.

The global network state as a theoretical framework

It is the intention of this book to employ the global network state (GNS) developmental model as a framework from which to examine China's state capacity at the sectoral level. In contemporary literature, a latecomer nation state that possesses the ability to transform its economic policy prescriptions and tools to respond to global economic challenges and opportunities is termed a developmental state. Developed as a response to the failure of both neoclassical economics and dependency theory in explaining the rapid economic growth experienced in East Asia, the developmental state model has become the dominant paradigm for understanding East Asian economic development, in particular, with the post–World War II Japanese economic model (Johnson, 1982, 1995).

A developmental state is defined here as a state that is actively involved in economic development policies beyond the normal functions of providing public goods and protecting social and national interests (Evans, 1995:45–48). To be developmental, a state needs to possess two critical and interrelated characteristics: state capacity, which requires a well-developed and coherent state apparatus and state-owned enterprises to implement state policies; and state autonomy, which involves the relationship between the state and societal, as well as extra-societal (or foreign), pressures. Each characteristic is necessary but not a sufficient condition for effective state intervention (Weiss, 1998). Weiss (1998) posited that:

> the relative success of states adjusting to changes in the international economic architecture is dependent upon, in part, their transformative capacity. Central to this capacity is a sufficiently robust partnership between relevant state actors, economic sectors and civil society actors such that policy changes can be negotiated and implemented. Weiss asserts that states are not unitary monolithic structures but are instead "organisational complexes" whose "parts" represent different ages, functions and (at times) orientations. (Weiss, 1998:15–16)

From this frame of reference, strategies for industrial development are not formulated and implemented by the state alone, but through a complex network of policy linkages, bureaucratic agencies and sectoral interests, a form of governance Weiss terms "governed interdependence" (Weiss, 1998:38).

However, whilst it is the intention of this book to argue that technological and industrial upgrading require specific sets of coordinated government interventions in both the domestic and international economy, I want to emphasise that changes in the structure of the global economy and the markets that comprise it, mean that it is not theoretically prudent to construct a state–market dichotomy as a theoretical lens.

To the contrary, Peter Evans argues that the predisposition of theorists to frame arguments about state intervention, in terms of state versus the market, need to be replaced with arguments that endeavour to examine exactly what kinds of state intervention are effective and their effects (Evans, 1995:10). In the state–market dichotomy where the state and the market are treated as being in a constant-sum relationship, the theoretical attention focuses on the degrees of departure from ideal-typical competitive markets. Conversely, the state either wanes or persists. However, asserts Evans, in the contemporary world, state withdrawal or involvement are not alternatives. State involvement is a given (Evans, 1995). Hence, the crucial question that theorists need to be addressing at this point is not how much state involvement is appropriate, but rather what kind and at what level?

Furthermore, extant studies of state capacity have essentially neglected to consider how the interventionist state has evolved with the growth of markets. The key question that needs to be posed at this point is: as the validity of the traditional state-centric developmental model loses its pertinence due to external and internal pressures for economic openness, liberalisation, increased deregulation, and structural reform of domestic market institutions, how exactly does the interventionist state endeavour to redefine and transform its economic apparatus and coordination agencies?

Moreover, what is important to note here is that the PRC sits uncomfortably in the wider literature on the developmental state. Firstly, China remains a one-party authoritarian regime. The Chinese polity is a state-centred system in which the patriarch state has the ultimate power and authority over society (Bell and Feng, 2007:62). As a direct result, the policy-making process is characterised by a system of elite politics whereby the leading participants in the policy process are individual leaders and top bureaucrats (Zhao cited in Bell and Feng, 2007:62). This elite hold on power is reinforced by a lack of organised interests between industry and the public. As a direct result, economic policy making in China is defined by an elite-led, top-down, decision-making process in which societal and economic actors that do not fit into the bureaucratic agenda are either ignored or marginalised (Bell and Feng, 2007:62).

However, it is important to note here that decision making in China is also subjected to political contestation and bargaining. For example, Peter Lovelock and John Ure (1998) use a two-tier bargaining model in order to construct a political economy model of bargaining. It is a theoretical framework that still has empirical and theoretical validity in a contemporary context. Essentially, Lovelock and Ure argue that in China there exists a recognised set of international policy objectives and a recognised set of domestic policy objectives that determine the parameters of the bargaining and negotiating framework. From this perspective, interested actors can bargain to achieve their objectives at either level, but the government's position as the arbiter between the two tiers is, asserts Lovelock, what has allowed it to preserve its coordinating role. Essentially, the picture that emerges from Lovelock and Ure's analysis is a complex bargaining process where policy implementation is slow and incremental. However, it is important to note that in a contemporary context decision making is not always as protracted as it was in the earlier stages of the development process, as relational ties have often already been formed between Chinese and international actors and the necessities and pace of technological development in the wireless communication sector require that actors reach decisions more quickly in order to meet market objectives and leapfrog their competitors.

In Lovelock and Ure's framework the

> "upper tier" of the mode, represents China's relation with the outside world which is motivated by a set of outward- (export-) oriented growth objectives and inward-oriented development requirements. Policy at the "lower tier" – or domestic level – is dominated by the government and/or Party technocrats, who are essentially unconstrained by independent law or institutions. Below this elite layer is a highly complex organisation of power that "fragments" authority along vertical/functional and horizontal/territorial lines. The net result is incessant bargaining and consensus-building among officials at all levels of the national hierarchy. (Lovelock and Ure, 1998:9)

Lovelock and Ure assert that in terms of policy making, Beijing has continued to utilise this two-level bargaining approach to coordinate developments: "Beijing has become particularly apt at switching negotiations from one tier to another in order to play vested interests off against each other" (Lovelock and Ure, 1998:9). Thus, although foreign partners specifically bring much-needed capital to the Chinese telecommunications development program, they do not negotiate over capital at the state level – where such concessions (as to foreign direct investment) must be made. Rather they negotiate over market access, market share, technology transfer, etc. Financial issues are negotiated at the locality – as part of specific domestic projects – where the defining regulation is that foreigners can neither own nor operate telecommunications networks. Yet, as Lampton et al. (1992) have shown, the bargaining that has characterised China's interaction with the outside world in gaining access to high technology and industrial finance has been replicated domestically between contending bureaucratic and entrepreneurial interests (Lovelock and Ure, 1998:9).

Moreover, Lovelock and Ure conjecture that whilst these two tiers are intertwined, they also contain distinctive characteristics. Hence, foreign firms will often need to bargain at each level over the same issues or same resources. Furthermore, empirical evidence indicates that the Chinese government has adopted a strategic policy of switching negotiations from the international to the domestic level (or vice versa) in order to achieve its bargaining objectives. That is, asserts Lovelock, "on the international level we find the Chinese government bargaining with MNCs and International Organisations (IOs) but, on the domestic level, MNCs and IOs bargaining with the Chinese government" (Lovelock and Ure, 1998:8).

However, whilst this two-tier framework provides us with a medium from which to structure our analysis of the Chinese bureaucratic process and the complex interactions that occur between Chinese economic actors, international firms and the Chinese state, it does not provide a sophisticated enough mechanism capable of capturing the complexity of contemporary relational and technological interdependence and the ways in which it operates to constrain and define the bargaining power and objectives of both Chinese and international actors.

Firstly, we need to extend the framework to incorporate a third tier. This third tier is designed to capture the patterns of bargaining and policy making that occur multilaterally through multilateral institutions like IMF, the World Bank, and the World Trade Organisation (WTO). It is these negotiations that produce the macro rules or principles that govern FDI and the range of legitimate economic trade tools and institutional structures available for national developmental use. Secondly, I want to incorporate the concepts of network legitimacy, reciprocal independence, relational networks and collaborative governance as important contemporary variables that operate to mould and define the parameters of the contemporary alliance building and bargaining environment in the wireless communication sector.

Hence, due to the presence of an array of complex and multifaceted variables and bargaining behaviour, it is imperative that we frame our work with a comprehensive contextual understanding of the actual market and state sectoral-institutional tools we are examining. Current literature pertaining to the developmental state is primarily focused on economy-wide aggregate-level analysis (Weiss, 1998; Evans, 1995; Besson, 2004). However, markets, of course, like states, are not monolithic structures but diverse entities that are divided by technological, political and organisation variables.

Indeed, the contemporary literature pertaining to globalisation and state capacity is defined by its theoretical and deductive style of exposition (Moore, 2002:12). As a direct result, the conclusions about the impact of globalisation – and their associated expectations about the resilience of the institutional specificities of particular markets in the context of such forces – are largely informed by theoretical reflection (Moore, 2002:12). The problem here is that such analyses are devoid of detailed, substantive, empirical analysis of "actually existing" markets. In order to address this theoretical deficit, this book will be framed around an industry-specific analysis of the global wireless communication sector.

The Chinese developmental state and the global wireless communication sector

I have selected the wireless communication sector as a case study precisely because it provides us with a good illustration of the advantages that firms may gain through strategic alliance membership as well as provides a theoretical point of entry for the isolation of exactly what type of institutional capacities the Chinese state needs to devise in order to accommodate this emergent form of capitalist organisation. The substantive research at the empirical level is designed to examine both the nature of existing capacities as well as efforts by the Chinese state to forge new ones.

As this book will highlight, the key role of the Chinese government as a developmental state in the contemporary wireless technology sector is not just about fixing market failures, but rather about actively creating the market for the new technologies by envisioning the opportunity space and allowing the right network of private and public actors to meet in order for radical innovation to occur at a global level. This has entailed a shift to what Sean O'Riain terms, a "flexible development state" (FDS) or "global network state" (GNS). A GNS, asserts O'Riain, is defined by its capability to form networks for the production of technological innovation, and by its ability to attract foreign investment and stimulate domestic growth. It achieves these key goals by providing information about transnational corporations and world markets to domestic firms, and develops institutional frameworks and funding decisions designed to provide the necessary conditions for stimulating connections between domestic and international companies. That is, the developmental network state's overarching goal is to facilitate collaboration and encourage innovation at both a domestic and global level (O'Riain, 2000). The key difference between earlier forms of the developmental states, argues O'Riain, is the way in which it is not just focused on the development of national champions and sectoral pillars but also on the construction and management of transnationally integrated local networks and markets (O'Riain, 2000:165).

Analytical framework and key concepts

Definitions of technological change and systems adaptation are inherently messy. This is a direct product of the fact that the processes that define technological development and change are comprised of a plethora of interrelated variables that coexist in a multilevel, dynamic and increasingly spatially diverse socio-political technological environment. In order to provide a relational lens capable of isolating the diverse sets of highly interdependent and conflicting theoretical variables that define the innovation and technological development processes of nation states in the contemporary, globalised socio-technological system, this book will be framed by the theory of complex adaptive systems (CAS). The reason for this is that CAS theory provides an important medium that helps us to understand the changes that are occurring in the wireless socio-technological system and the way they are both impacting on and enabling state and firm responses to this change. Without a comprehensive understanding of the "actual mechanisms" of socio-technological systems change, it is not possible to hypothesis about the way in which technological change and the shift to an innovation-based collaborative economy has occurred or has both shaped and empowered system actors' behaviour.

Globalised socio-technological systems are high-impact, hard-to-understand, technology-intensive systems that possess significant societal, political and economic implications (Sussmam, 2012). CAS as a theoretical framework will be employed throughout this book to explore changes in both the global and the Chinese socio-technological system and to locate the structural pressures, dependencies, networks and agency opportunities that the emerging global socio-technological systems exert on Chinese state leaders and their technology development agenda.

CAS theory helps us to model systems change and strategic planning and agency by providing a medium from which to examine and conceptualise the way in which governments and firms are adapting to socio-technological change and the globalisation of technological development and production.

A complex adaptive system is comprised of a complex network of adaptive agents. Complexity as a term denotes the density of causal interconnectedness within a system. These connections are defined by complex systems of elements that are interdependent and interconnected by multiple feedback processes. Primarily, complex technological systems are both a "scientific knowledge" and an "economic policy" system with flows of resources and information taking place among its component nodes and across its boundaries. The resource flows include finance, policy decisions, materials, human capital and technological inputs. The knowledge flows include formal and tacit knowledge. They also include "learning" about how to scale out technology at both a centralised and decentralised level. Such an innovation system is classified as evolutionary since new knowledge is constantly entering the system and facilitating behaviour modifications. There is no return to a previous equilibrium. They exhibit complexity in that knowledge and resource flows are moving across many stakeholder groups. This requires system actors to minimise and manage complexity. The systems themselves are primarily adaptive and resilient while resources flow across their boundaries. They behave holistically as a totality and therefore, analytically, their behaviour cannot be reduced entirely to that of component nodes. Finally, they engage in networking in order to facilitate information interactivity and improve system efficiency (Halle and Clark, 2009).

Complex systems theory highlights how connectivity and interaction between system actors involves the selection of linkages and the prioritisation of relationships that are established with other systems. Specifically, this ability is what defines the degree of openness of the system at different levels of aggregation. Thus, an innovation system depends on the "system making" connections between the components of the relevant ecologies that ensure the flow of information directed at innovation problems. Hence, there is a close connection between the notion of trajectories of technological solutions within a particular technological paradigm, the evolving problem sequence, and the dynamic notion of an innovation system (Dosi, 1982).

This ecological inheritance cannot be understood without reference to the inherited technological and political structures that both constrain and enable socio-technological evolution and development. Bell (2011) argues that in order to understand actor-centred agency we need to develop a framework that considers the importance of interpretative agency and bounded discretion within institutional and structural settings. Similarly, in order to fully understand the evolution of technological change and its relationship to structural features in the socio-technological environment, we need to conceptualise the evolution of technology as an inherently multilevel process (Tushman and Nelson, 1990); that is, historically contingent with subsequent inventions building upon one another. Hence, an ecological framework for studying technological change examines technological evolution in terms of its elemental components and the ways in which they recombine over time. From this theoretical perspective we need to model agents both as partially constrained by their immediate institutional contexts and also as operating in institutional and structural settings that constantly evolve and potentially open up new opportunities for agents. Furthermore, structural contexts, according to Bell, include the broader political, economic or social environments that operate in a "strategically selective" manner, establishing incentives or disincentives or other rationales for action that may lead agents to favour certain developments or choices over others. As Bell asserts: "Although structuralist's accounts are often handled in a way which implies deterministic patterns of constraint, in reality, structures can both help constrain and empower agents. The pattern of constraint or enablement embedded in such relationships can also change systematically over time" (Bell, 2011).

The ability of system actors to respond to complex technological systems change can be measured by examining their policy adaptions, network connections, new forms of innovation capacity, and the development of a system of multiple nodes of expertise, such as new technological products, policy directives and services which have become prominent nodes in their own right (Halle and Clark, 2009).

Research methodology and data collection

As it has already been established, the empirical study concentrates on the global wireless communication sector as an appropriate industry to study Chinese state capacity, competitive and technological upgrading, and alliance capitalism. Specifically, I am interested in examining the institutional, organisational, relational and policy adjustments implemented by the Chinese government and its key domestic firms in response to the dynamic technological and regulatory environment that the global wireless communication sector occupies. The primary empirical focus is on the transition from 3G to 4G technologies with the shift from a hierarchal global and domestic governance structure and proprietorial knowledge ownership systems to that of a more horizontally integrated global alliance architecture and network structure and a more open and shared platform for global technological development and knowledge sharing.

In order to empirically examine Chinese state leaders' strategic attempts to intervene in the economy and develop a set of policies to facilitate the development of dynamic capabilities, we need to examine how the various interests of the state, local actors and the global economy are represented and coordinated by the government of China. We need to ascertain how and in what ways the state is embedded in professional-led networks of innovation and international capital, and whether or not its organisational structure is flexible enough to effectively manage the complex and diverse interests it is presented with. We also need to examine the mechanisms that the state is employing to create an economic environment where domestic firms can upgrade their technological and organisational capabilities and hence increase their indigenous knowledge production and technological development. Furthermore, we need to explore what strategies the Chinese government is employing to embed global firms in its economy in new ways that foster reciprocal interdependence as opposed to the shallow integration and technological dependency produced by previous failed technology development strategies such as techno-nationalism.

Obviously, dynamic firm and state capacity development, market building and the construction of international alliances are non-linear processes, devoid of any deterministic order in their development. This of course means that isolating variables for analysis is a very difficult task. However, key empirical variables that can be isolated for analyses for the purposes of this book are: elite political control over the development process; relational networks, either state, managerial or entrepreneurial, for establishing dynamic capabilities and positioning products in domestic and foreign markets; indigenous and collaborative standard-making attempts; the occurrence of ecosystem building alliances; network legitimation; relational and structural reorganisation of the state's architecture in the sector and policy directly relating to the regulation; and incentive structure pertaining to both domestic and foreign investment in the sector.

Furthermore, because I am interested in how Chinese state leaders are targeting global alliance ecosystem building and cooperative R&D in a bid to insert Chinese wireless firms at the cutting edge of the high-technology productivity frontier, we need to empirically examine the institutional, regulatory and bargaining strategies that the government of China is engaging in to gain global leverage and legitimacy. From an empirical perspective, this requires us to examine the wider bargaining frameworks, broader sectoral agreements and terms of engagement that occur between global wireless technology firms, business associations and standards bodies and the political interests of Chinese state leaders. What I am interested in here is isolating and examining both firm and state adjustments to achieve political and economic leverage in a global innovation-based market structure that is becoming increasingly collaborative and structurally interdependent.

In order to achieve this goal, data on governmental–business relations and alliance-bargaining strategies has been collected from both primary and secondary sources. Because no comprehensive database exists for the identification of international R&D alliances between Chinese and foreign firms in the global wireless communication sector, a new database was developed for this book. The primary data was collected from company annual reports, governmental policy documents, economic/market data and in-person interviews that involved the administration of a semi-structured questionnaire. Interviews were conducted with Chinese business and academic actors that are responsible for the development, implementation and management of strategic policy development and innovation in the wireless communication sector. The interviews were conducted in China in 2011 and 2012. There were 14 interviews, and the average duration of each interview was two hours. The interviews were subsequently documented and transcribed for analysis. The secondary data was drawn from multiple sources, including: journal articles, press releases, academic texts, corporate websites, newspapers (Lexis-Nexis database), magazine articles (Business Source Premier Database), Internet news, (3CPP2) and other public sources.

The alliance database developed for this book includes agreements that contain arrangements for the formation of strategic alliances between multinational corporations and Chinese firms, and universities and government organisations that specifically involve cooperation in research projects, joint product development, R&D joint ventures and R&D consortia. The focus of the data gathering was on those forms of cooperation and agreements for which joint R&D activities or ecosystem-building alliances are at least part of the agreement. My sample is comprised of R&D and ecosystem-building alliances from North America, Europe, Japan, South Korea and China. Included in the sample are Chinese telecommunication providers, wireless equipment manufacturers and key research institutes that engaged in R&D and ecosystem-building alliances within the 1999–2012 period.

Organisation of the book

This book is comprised of seven chapters, including the conclusion and the introduction. In this introductory chapter I have sought to introduce the main arguments and concepts of this book. Specifically, I have sought to introduce the idea of critical ecosystem dependencies and highlight how it has facilitated the emergence of alliance capitalism as an emerging technological development strategy. I have also outlined how this book will be framed by the theory of complex adaptive systems and the idea of socio-technological systems shaping and agency.

In Chapter 2, I will provide the reader with a contextual understanding of the globalisation process and the early external constraints it placed on China's developmental strategy. The chapter will examine the process of FDI as a development strategy and outline the key reasons why it has failed to generate significant technological development. Chapter 2 will also outline recent failed techno-nationalistic attempts to construct an indigenous development agenda in China. It will illustrate how foreign interests and the emergence of a more collaborative economy constrained China's attempt to adopt a "go-it-alone, techno-nationalistic" development strategy using wireless LAN authentication and privacy infrastructure (WAPI) as an example.

In Chapter 3, I will outline a framework for the examination of high-technology innovation policy in an environment defined by complex global systems change, critical ecosystems dependencies and the emergence of technical co-development and collaborative value appropriation. I will argue that in order to respond to socio-technological systems change the government of China is currently engaged in an extensive overhaul of its technological development strategy. This time it is replacing its "go-it-alone, techno-nationalistic" strategy with a globalised adaptive strategy that is focused on the pursuit of strategic agency and global ecosystem embeddedness.

Chapter 4 is designed to develop a theoretical and empirical overview of state capitalism as a developmental model for globalised, innovative upscaling. The developmental state as a theoretical model for economic and technology development will be examined here. It will highlight how the Chinese state understands and responds to socio-technological innovation, examines its ability to develop indigenous knowledge assets, and its capacity to insert itself and its domestic firm into emerging technological ecosystems in order to facilitate global technological alliances in the emerging collaborative economy.

In Chapter 5, the institutional, regulatory and bargaining strategies that the Chinese government is utilising in the 4G-TDD global wireless communication sector in order to gain the global leverage and legitimacy necessary for global alliance ecosystem building and cooperative R&D will be examined. Specifically, it will examine how China's new, more alliance-based, high-technology development strategy combines the development of core indigenous technological assets with the development of collaborative R&D centres and socio-technological alliances with foreign firms in order to facilitate the development of collaborative dependencies and global network embeddedness. It will highlight the way in which this globalised alliance-based strategy responds to contemporary socio-technological systems change and operates to generate the relational ties and collaborative alliances that are essential for the development and appropriation of technological knowledge assets and commercialisation processes.

Chapter 6 is empirical in nature. It focuses on the global wireless communication sector, presenting empirical evidence that can support the argument advanced by this book that due to socio-technological systems change and the need for product convergence and interoperability a new form of "state-led" alliance capitalism is emerging in China. Its primary focus will be on the global positioning of Chinese domestic firms in global innovation and production networks and the occurrence, nature and number of strategic R&D alliances that exist between Chinese and foreign firms in the sector. The data presented will highlight the increasing innovative capacity of key Chinese firms operating in the sector and highlight how these firms have managed to become firmly embedded in the contemporary and future global wireless communication sector's ecosystem through the development of innovative core technologies, the co-shaping of intellectual property standards and a focus on collaborative global innovation network development.

The concluding chapter is designed to provide the reader with a summary of the main arguments presented in the book in relation to the emergence of alliance capitalism and its impact on the Chinese government's technological development strategy. It will make suggestions for further research. Specifically, there is the need for researchers to position any analysis of Chinese technological and economic development within a framework that acknowledges the increasing relational and interdependent nature of contemporary high-technology development as the emerging collaborative economy, and the increasing existence of critical technological dependencies fundamentally redefines the spatial and relational parameters of the global economy.

2

The Perils of Strategic Technological Development Policy: Two Failed Chinese Attempts, FDI and Techno-Nationalism

Until recently, the Chinese approach to industrial and technological development focused on attracting foreign direct investment (FDI) to leapfrog the economy. FDI is a key development strategy utilised by developing countries with minimal capital reserves using cheap labour as a key resource. It facilitates the importation of capital, equipment and technology in order to build an industrial base and generate capital reserves from exports (Thun, 2006:3). Between 1985 and 2005, it is estimated that the annual net FDI inflows into China grew from US$1 billion to US$72 billion. In addition, it is estimated that within this period China absorbed more than US$600 billion in FDI. This is a figure that is 12 times higher than the total stock of FDI Japan received between 1945 and 2000. In the early 1990s, Beijing approved a new form of enterprise termed wholly foreign-owned enterprises (WFOEs). By the early 2000s, WFOEs attracted 65% of new FDI in China. Furthermore, since 1993, China has become the largest recipient of FDI among developing countries (Pan, 2009:16).

It is important to note that the form of this investment is characterised by commitments to new assets. This contrasts with the inward investment of other leading investment destinations, such as Luxembourg, which primarily attract financial transactions, whilst large OECD countries, such as France, Germany, the United States and the United Kingdom, primarily involve mergers and acquisitions; that is, the trading of existing assets rather than the development of new assets (Enright, Scott, and Chang, 2005:2). Indeed, institutions such as the World Bank have credited FDI as the major driving force of China's economic success (Zheng, 2005:19). Academic researchers tout the enormous benefits of FDI for China citing the benefits of technology and knowledge transfers, the introduction of market behaviour and capital infusion (Zheng, 2005:19). Unfortunately, a plethora of empirical evidence also reveals FDI has played only a limited role in transferring technology to key Chinese technological sectors and has failed to contribute to local technological and innovative capability-building processes (Young and Lan, 1997; OECD, 2002; Sigurdson, 2002; Gallagher, 2003; Sjoholm and Lundin, 2013).

As a direct result, the Chinese government during the early 2000s endeavoured to adjust its earlier development strategy from one that focused on export-led growth and the importance of FDI, to the development of a go-it-alone techno-nationalistic strategy and a desire to increase domestic consumption. Operating from the assumption that the size of the Chinese market – with 1.3 billion potential consumers – would provide a powerful bargaining tool, the Chinese government endeavoured to construct a policy paradigm structured around the idea of indigenous innovation and the proprietorial ownership of standards and intellectual property rights. However, the Chinese government was forced to abandon this new techno-nationalistic strategy in its very early stages as the international community and its multilateral institutional bodies quickly reacted to contain China's techno-nationalistic ambitions.

It is the intention of this chapter to examine China's early focus on FDI as a development strategy and outline the key reasons why it has essentially failed. It will also outline recent failed techno-nationalistic attempts to construct an indigenous development agenda. It will illustrate how foreign interests and the emergence of a more collaborative, alliance-based global economy have played a major role in constraining China's go-it-alone strategy using wireless LAN authentication and privacy infrastructure (WAPI) as an example.

FDI: trading market access for technology

Prior to 1979, FDI was prohibited in China. However, in 1978, the Chinese government, employing a mercantilist strategy, began to open up the Chinese economy to transnational economic actors. In a strategy that has been called trading market access for technology, the Chinese government sought to utilise the size of its market and abundance of cheap labour as an incentive to attract the world's major investors to transfer technology in order to modernise the Chinese economy.

In a dual-track strategy that has been termed "growing out of the plan", Barry Naughton (2005) argues the Chinese government sought to experiment with new ways of organising economic activity without dismantling the command system until successful results had been created in the experimental production zones. This included new at-the-border institutions – channels of global transaction through which all foreign transactions had to flow. Specifically, these included special economic zones (SEZs), high-tech or export-processing zones, joint ventures (JVs), and a number of foreign affairs offices with various objectives (Zweig, 2002:23). In these special economic zones, economic actors were granted regulatory powers and preferential privileges by the state that were designed to encourage economic exchange and decrease the transaction costs of international transactions (Naughton, 1995). In this way, a new political economy arose with an incentive structure designed to redefine the economic behaviour and interaction of bureaucrats, local officials, ordinary citizens, enterprise managers, collectives and foreign investors (Zweig, 2002:23).

China's policy was clearly designed to encourage export-oriented FDI, looking externally to draw on both inputs and markets, and granting well-defined freedoms and incentives to the foreign-invested enterprise (FIE) sector. Policymakers, by developing the coastal development strategy that afforded SEZ-like privileges to the entire coast of China, created a kind of gigantic export-processing zone, where free markets were defined not so much by geography but rather by ownership (Naughton, 1995:302).

As a direct result of these governmental initiatives, FDI inflows into China increased rapidly. In 1984, a new foreign investment law further accelerated China's inward FDI growth, with another sharp increase occurring after 1992 when China reaffirmed the policies of openness and market-oriented reforms. The massive inflows of FDI within this period were representative of a growing trend for transnational corporations (TNCs) to outsource and subcontract production and even services to China. According to a 2006 World Investment Report, China ranked highly as one of the most-favoured locations of both the world's largest TNCs and the largest TNCs from developing countries. In a contemporary context, corporations from 190 countries and regions, including 450 of the global top Fortune 500 multinational corporations, have invested in China. Moreover, a further 60,000 foreign-owned factories opened in the period between 2003 and 2005 (Pan, 2009:16). Whilst it must be granted that this FDI helped provide a medium for Chinese enterprises to become extensively linked with the global economy, this industrialisation was itself "shallow". Economic development necessarily implies structural transformation, primarily, from low productivity to high productivity activities, from agriculture and simple manufacturing to modern industry (Steinfeld, 2004:253). The "Japan model" of developing countries progressing up the technology value chain depicts seven stages of developmental sophistication that developing nations move through. These include: (1) light industry, (2) assembling and processing, (3) expansion into heavy industry, (4) increases in local content, the employment of more technologically intensive processes and the investment in brand-name development, (5) top-shelf electronics, export of capital intensive goods and domestic market drives growth, (6) world leader in the production of high-technology goods but not involved in indigenous innovation processes and (7) innovators in technology and the knowledge economy. However, a number of theoretical and empirical studies indicate that in the contemporary international economy movement beyond the third stage is problematic (Steinfeld, 2004:253). The overarching problem here is that as a nation's growth slows, it becomes stuck in what has been termed the "middle-income trap" from which it cannot escape (Aiyar et al., 2013).

The term "middle-income trap" was first advanced by Gill and Kharas (2007) in a bid to describe the occurrence of slowdowns in a number of former East Asian miracle economies. Gill and Kharas proceeded to conjecture that the ability to sustain growth through the middle-income band required substantial reforms to the institutions of economic policy making and political processes (Gill and Kharas, 2007; Yusuf and Nabeshima, 2009; Woo, 2009; Ohno, 2010; Resen, 2011 cited in Robertson and Ye, 2013).

Indeed, historical data indicates that very few developing economies have managed to graduate into a high-income nation in the past 50 years. Countries that have made the transition successfully include Japan, Israel, South Korea and Singapore. A 2012 study conducted by the Asian Development Bank conjectured that in 2010, 35 out of the 52 middle-income countries the author studied around the world were trapped. Another study by Barry Eichengreen (an economist at the University of California, Berkeley), Kwanho Shin (at Korea University in Seoul) and Donghyun Park (at the Asian Development Bank in Manila) found that growth can slow down quite precipitously in countries that experience the trap (Schuman, 2011).

It has been conjectured that China's growth will soon slow down and it will also become a victim of the "middle-income trap". For example, Acemoglu and Robinson (2012) argue that: "China's extractive political institution is not compatible with innovation and high growth in the long run. In their view, although the growth process driven by catch-up, import of foreign technology, and export of low-end manufacturing products may continue for a while, it is deemed to come to a halt as soon as China reaches the living standards of a middle-income country" (Wang, 2013).

Indeed, China's developmental strategy in this period, with its overt focus on FDI, was not able to move domestic firms' activities beyond that of non-differentiated commodity production or allow any degree of control over their position in global production networks. This problem, it has been argued, is partly a reflection of China's nationalist industrial policies. This is because China's "pillar" industrial strategies, which were structured around the political ideas of self-sufficiency, operated from the assertion that a country can still create a particular industry from upstream to downstream and the whole supply chain can operate within its own territory. What these strategies did was to artificially force the local vertical integration of industries and clustered activities, regardless of technological connectivity, under the confines of a select number of national group enterprises. Furthermore, it was a strategy devised to reduce China's dependence on the foreign importation of key components by facilitating the development of a domestic supply chain. However, the problem was that because the state failed to comprehend the dynamic development of emerging, collaborative, global innovation networks and technological ecosystems its intervention in the market to assist in the development of its national industries essentially failed.

It is important to note here that China did not have a coherent predetermined technology plan in this period. As Naughton and Segal (2003) articulated:

> the future is likely to display the continuing salience of techno-nationalist attitudes, but without the coherent, tightly integrated policy package that supported techno-nationalism in Japan during the 1950s through 1970s. China will select bits and pieces of preferential policies, designed to advance techno-nationalist ideals within the context of a fiercely competitive and fairly open domestic economy. Such policies will often seem to lack intellectual coherence, and represent purely adaptive, opportunistic policies of "muddling through." In the past, though, "muddling through" has been a fairly effective approach for policy-makers trying to cope with China's diversity and dynamism. Perhaps it will be in the future as well.

In the following section I will provide a brief overview of the Chinese government's key technological policy programs in this period.

China's early science and technology policy and programs

In the 1980s, the Chinese government designed a series of programs to accelerate its scientific and technological development. These include: the Key Technologies R&D Program, the 863 Program, the 973 Program, the Spark Program and the Torch Program (Swissnex, 2009).

Key technologies R&D program

The Key Technologies R&D Program, launched in 1982, was the biggest scientific and technological development program in China of the twentieth century. Oriented towards national economic construction, its aim was to solve critical, direction-related and comprehensive problems in national economic and social development, including agriculture, electronic information, energy resources, transportation, materials, resources exploration, environmental protection, medical and health care, and other fields. This program, engaging tens of thousands of researchers in over 1000 scientific research institutions nationwide, has had the largest funding, employed the most people and had the greatest impact on the national economy of any plan to date (Swissnex, 2009).

863 Program

In March 1986, the National Hi-tech R&D Program (863 Program) was launched after exhaustive examination by scientific experts. The program set 20 themes in biology, space flight, information, laser, automation, energy, new material and oceanography. The government's role here was/is one of macro control and support. The general direction of research is decided by scientists after discussion – and specific projects are decided by a committee of experts responsible for keeping abreast of international research and reporting annually on their own fields, so as to set new research directions. Another feature of the program is that its results can be quickly industrialised (Swissnex, 2009).

973 Program

The 973 Program was launched in 1998 as a key national program for development. It primarily involved multidisciplinary, comprehensive research on important scientific issues in fields such as "agriculture, energy, information, environment of resources, population and health, and materials, providing theoretical basis and scientific foundation for solving problems. The program encourages outstanding scientists to carry out key basic research in cutting-edge science and important sci-tech issues in fields with a great bearing on economic and social development. Representing China's national goals, it aims to provide strong scientific and technological support for significant issues in China's economic and social development in the 21st century" (Swissnex, 2009).

Torch program

Launched in August 1988, the Torch Program is one of China's most significant early hi-tech industry programs. Its overarching goal was to provide national guidelines for science and technology (S&T) policy development. The program focused on projects in new technology fields, including new materials, biotechnology, electronic information, integrated mechanical-electrical technology, and advanced and energy-saving technology. It also focused on the creation of high-technology development zones. Here, the intention of Chinese policy makers was to recreate the experiences of Silicon Valley, Route 128, and other science and technology parks locally by locating universities and high-technology firms in the same area and combining research and education with production. In order to achieve these fundamental goals, it offered preferential policies in five key areas to firms operating in the parks, including: taxes, finance, imports and exports, pricing and personnel policy directives (Swissnex, 2009).

Spark program

The Spark Program was implemented in 1986. Its overarching goal was to revitalise the rural economy via the development and popularisation of S&T in rural areas in order to improve the lives of the rural population. It resulted in more than 140,000 S&T demonstration projects being carried out in 90% of rural areas throughout China (Swissnex, 2009).

These policy programs, it is argued, represent an attempt by Chinese policymakers to replicate the earlier techno-nationalistic success of Japan and Korea. Here the primary techno-nationalistic agents were to be large SOEs and government research institutes. Where MNCs were to play a role, the intention was that they were to be partnered with strong domestic SOEs (Naughton and Segal, 2003).

However, in contrast to the earlier success of the techno-nationalistic policies pursued by Korea and Japan, it became increasingly clear that the techno-nationalistic policies devised by the Chinese government were not suited to the emerging economic environment. For example, by the late 1980s, the 863 Program had only achieved limited success in bringing new products to market. Moreover, participating research institutions had few official connections with enterprises, and these enterprises had few incentives to cooperate with these institutions to develop new innovations. Moreover, it has been asserted that the Chinese government was too short-sighted and sought to achieve significant results with limited resources in too short a period of time. For instance, from 1988 to 1994, the average research fund for 863 Program researchers was only US$5000. This is because the funds were spread on average over 1044 research programs annually (Naughton and Segal, 2003).

In contrast, the Torch Program had a degree of success in commercialising technologies and supporting the growth of high-technology industries. However, there existed too many high-technology parks throughout the country with not enough enterprises to sustain them. Moreover many enterprises did not apply for the preferential funding available and failed to innovate at all. Conversely, by the early 1990s, these policy programs were being subjected to increasing scrutiny and a new, more rigid form of techno-nationalism based on the idea of indigenous innovation, and standards began to emerge (Naughton and Segal, 2003).

It is important to note at this point: "the overall pattern of technology development during the reform period has been one of restless change" (Naughton and Segal, 2003). This is a product of the fact that the size and diversity of China allows for more policy experimentation, argue Naughton and Segal (2003). Hence, it has the option of maintaining two or more separate, competing, and not necessarily integrated, approaches towards technology acquisition (Naughton and Segal, 2003). The results are a national mosaic that varies both temporally and geographically. Thus, whilst in the following section I will examine Chinese policymakers' attempt to facilitate indigenous innovation and standards development, it is important to remember that the experimental, fragmented and changing nature of Chinese developmental policy making demands an awareness of the conflicting local and national political and economic objectives that exist between diverse economic and political actors, and thus the need to draw multidimensional and nuanced conclusions.

Shallow integration and the liberal critique

The inability of the Chinese government's FDI strategy and its S&T programs to move its firms up the high-technology value ladder in this period, and the lack of technological spillovers from FDI, lead to increasing concern and debate in China over its low value-added position in global economy. Chinese intellectuals can essentially be divided into two loose ideological categories, the Liberals and the New Left. The Liberals essentially support the theoretical propositions of the free market. In contrast, the New Left, who are essentially techno-nationalists or mercantilists, are critical of neoliberal policy in its various guises, whether as a rubric for free-market economics or as a broader metaphor for Western interpretations of modernity (Fewsmith, 2007:1). Essentially the New Left assert that emergent economies such as China can only survive through the development of economic policies that protect national industries against foreign competition and through the exploration of overseas markets for domestic accumulation (Bell and Feng, 2007:58). From this perspective, foreign investors and MNCs are predatory and only come to China for profit (Bell and Feng, 2007:58). They assert that foreign companies have little intention of transferring their core-advanced technology to China, which they view as a strategic competitor (Bell and Feng, 2007:58).

Hence, according to this view, the Chinese government should focus on indigenous investment and economic development in order to achieve economic independence. Furthermore, it is argued that China's huge market of 1.3 billion people provides an important point of political and economic leverage that that could be utilised to force foreign companies to comply with China's own de facto technical standards (Bell and Feng, 2007:59). Alarmist domestic critics of China's reform-era development strategy have gone further than government officials to declare the failure of the entire neoliberal economic orthodoxy behind the "comparative advantage" and "trading markets for technology" doctrines. Zhong Qing, for example, noted that while markets have been traded out, technologies have not been brought in. Indeed, academics and policymakers within this period became increasingly critical about the presence and behaviour of foreign firms in China, asserting that these firms charge unduly high licenses for their patents, "crowd out" domestic firms in the market for highly skilled labour, monopolise technology standards, and thwart technology transfer and knowledge spillovers (Zhao, cited in Pan, 2009). As former Commerce Minister Bo Xilai articulates, China's position in the global economy can be compared to "trading 800 million shirts for one A380 airbus" (Zhao, 2008).

In October 2005, this discourse on indigenous technological development was extended further in a number of important central party state documents, including the October 2005 Communist Party of China (CCP) Central Committee Proposals on the 11th Five-Year Plan for National Economic and Social Development, and the National Informatization Development Strategy (Year 2006–20) approved by the State Informatization Leading Group (Zhao, cited in Pan, 2009).

It is also interesting to note that at this time the Chinese government erected a number of barriers to FDI. For example, whilst official government figures indicate that FDI in China rose by more than 13% in 2007, the European Union reported that its investment in China had plunged from approximately $7.9 billion in 2006 to around $1.5 billion in 2007 (Pan, 2009). The existence of these discrepancies, it has been asserted, is a result of the fact that the official FDI figures were driven by funds reappropriated by domestic enterprises through Hong Kong and offshore capital centres. Furthermore, this lack of genuine FDI, it is argued, is no accident. Instead, it represented a deliberate attempt by the Chinese government to restrict market access to foreign firms. Indeed, policy documentation does indicate that China's mercantilist tendencies intensified sharply in the fall of 2005, which can be seen reflected in the discussion of the sale of minority shares in state banks at the plenary meeting of the CCP Central Committee. Following this, the path-breaking acquisition in October 2005 of the state-owned Xuzhou Construction Machinery Group (XCMG) by the Carlyle Group, a US private equity firm, was reversed. Several sales that had previously been approved were vetoed at the March 2006 meeting of the National People's Congress (NPC). Additional industries were designated as "strategic" and thus made off-limits to foreign investors. During the CCP's plenary meeting in the fall of 2006, this limitation morphed into an outright ban on any type of FDI that threatened "economic security" (Zhang, 2008; Scissors, 2009; Kwok-wah, 2012; Zhihong, 2006).

However, a number of factors have undermined the Chinese government's ability to operate from a mercantilist economic stance and exploit its presumptive bargaining power relative to TNCs. The problem is both organisational and monopolistic. Foreign TNCs that invest in China at the time had a number of important ownership advantages. They created and controlled intellectual property and key technology standards and hence had become market makers, controlling the pace of innovation and decisively shaping the trajectory of their respective industries (Bach et al., 2006). This ownership of intellectual property rights allowed them to set the agenda, at an international level, and influence the way in which technology progressed; whilst their world-class brand names enabled them to gain direct access to customers and marketplaces, which in turn facilitated their initiation of concepts for product development and the means of further exploiting market potential elsewhere (Chen, 2004). The strengthening of international property rights within this period represented attempts by foreign firms to ensure that the technology and knowledge that provided them with this competitive advantage via architectural organisational control was protected at an international level. This operated to inhibit high-technology upscaling and value appropriation by domestic Chinese firms.

Essentially, in this time frame, Chinese firms operated in sectors that were dominated by foreign firms who monopolised international property rights. Indeed, it has been asserted that such high levels of foreign dominance over the production process itself requires us to question whether Chinese businesses were actually "Chinese" businesses at all. This is because in the export sector, firms that are often termed "Chinese" businesses are in actual fact Chinese subsidiaries of global multinational companies, and Chinese joint ventures with businesses from the industrialised enterprises are often WFOEs. For example, by the early 2000s, WFOEs accounted for 65% of new FDI in China. Moreover, it is important to note that it was these Chinese subsidiaries, or foreign-funded enterprises (FFEs), which accounted for approximately two-thirds of the total growth in Chinese exports within the period between 1994 and mid-2003 (Pan, 2009). In addition, foreign companies at this point in time managed virtually all intellectual property and accounted for 85% of the country's technology exports. As Gavin Heron, managing director of TBWA/Shanghai, has argued, China is "a story of international brands, not local ones... As soon as a local brand has any traction, they're bought out by a multinational" (Pan, 2009). Indeed, based on their technology and branding superiority, many foreign companies, it is argued, secured their supremacy in China's production (such as delivery dates, industry and quality standards, and design specifications) in this period without actual ownership over production. For example, through control of industry standards – a phenomenon known as "Wintelism" – Microsoft and Intel retain huge influence over access to the PC market without ever producing PCs themselves (Pan, 2009).

It is not surprising that the lack of a domestic technology base conspired to place Chinese companies in many industries at the mercy of their multinational counterparts, especially in terms of technology access. Indeed, it is estimated that between 60% and 80% of the value of all Chinese exports are processed (imported) components. Since the import content of the FFEs is often much higher, their exports from China yield still less value added for the "national economy" than the roughly equal value of exports from "national" firms. Thanks to Wintelism, leading foreign enterprises continued to control the realisation of value through their control of the sales channel and market standards. For example, Intel earned as much as 10% of its total US$30 billion a year in revenue from selling computer microprocessor chips to China. This point is as valid to labour-intensive products as it is to the high-tech sectors. A Barbie doll made in China is sold for US$20 in Western markets, but only about 35 cents is retained by the Chinese. In this way it can be seen that "Made in China" essentially means made by America in China or made by Europe in China, etc. (Pan, 2009).

Gilboy (2004) argues that China's technological advance is primarily a product of the dominant role of foreign investment, and China's domestically owned firms are defined by an "industrial strategic culture that encourages them to seek short-term profits... (and) forego investment in long-term technology development and diffusion....Most Chinese industrial firms... have not increased their commitment to developing new technologies. ...R&D expenditure as a percentage of value added at China's industrial firms is only about one percent, seven times less than the average in countries of the OECD" (Gilboy cited in Pan, 2009:43).

From this frame of reference, China's integration into global production networks has been at the expense of both the national coherence of and indigenous development of Chinese firms. Hence, the New Left assert that in a zealous drive to join with international standards (*yu guoji jiegui*), many Chinese companies have not only developed a dependence on transnational capital and technology, but more remarkably, some have deliberately avoided horizontal collaboration with their domestic counterparts, especially if such collaboration "crosses regional or bureaucratic boundaries" (Gilboy, 2004). As Gilboy points out: "China's best firms are among the least connected to domestic suppliers: for every $100 that state-owned electronics and telecom firms spend on technology imports, they spend only $1.20 on similar domestic goods" (Gilboy cited in Pan, 2009).

Moreover, it is argued that in this period, an examination of the "going global" strategy of many Chinese companies, which is often perceived as orchestrated by the state, reveals that they were not so much supported by an ambitious, monolithic state as they were "pushed" out by intense competition in domestic markets from FFEs. According to a 2003 survey of China's 50 largest "industry-leading" firms by the Shanghai office of the German-based Roland Berger Strategy Consultants, slightly more than half of the participating firms named "seeking new markets" as the overriding imperative for globalising their business activities. Among this group of firms, manufacturers cited, in particular, growing competitive pressure from multinational companies in the home market, excess capacity, and razor-thin and sliding profit margins as key reasons to search for new markets abroad. Thus, those Chinese companies going global, instead of representing a coherent national strategy, in fact, testifies to a weakening of their position in the domestic sphere (Roland Berger Strategy Consultants, 2003).

Furthermore, Wang (2006) identifies a dualistic pattern in China's technological development, with the export-oriented segments of the economy being relatively isolated from those producing mainly for the domestic market (Zeng and Wang 2007, cited in Wang, 2006) stress the weight of constraints, such as an insufficiently developed institutional framework, lack of indigenous intellectual property rights, relatively low overall educational attainments, the lack of a large pool of world-class talent, the embryonic stage of indigenous innovation capacity, and insufficiently developed linkages between R&D and industrial enterprises. Moreover, it is argued that the Chinese government failed to develop the necessary policies for encouraging technological cooperation between universities and industry, or the integration of the country's national science institutions into global innovation networks (Wang, 2006).

From this theoretical perspective, the Chinese government failed to develop a comprehensive FDI technological development strategy that could achieve both effective technology transfer from foreign technological leaders, as well as maintaining an appropriate balance between indigenous innovations and technology imports.

Indeed, the problem Chinese elites are faced with is that national governments that do not take into consideration international capital markets and foreign investors before devising economic policy objectives do so at their own risk (Thun, 2006:13). This is because nation states that fail to conform to the framework of supranational institutions – whether in the form of the European Union (EU), the World Trade Organisation (WTO) or the International Monetary Fund (IMF) in its contemporary form risk finding themselves locked out of the world trading system (Thun, 2006:13).

Indigenous development/techno-nationalism

In January 2006, the Chinese government launched the "National Medium and Long-Term Program for Scientific and Technological Development" (MLP) (2006–20). Operating from the assertion that FDI as a technological development strategy had essentially failed, Chinese state leaders introduced two new theoretical formulations into the political arena. These are the "scientific concept of development" and "indigenous development". These terms became powerful code words for China's reformulated development strategy and were frequently cited in state media and policy documents. In this long-term plan for S&T development, the most pertinent goal was to ensure that the Chinese nation state becomes a pre-eminent global economic and technological power relying on independent "indigenous innovation" (Gabriele and Haider, 2008:14).

In order to achieve this goal, the plan stipulated that: "By 2020, the nation's gross expenditures on R&D (GERD) are expected to rise to 2.5% or above of the gross domestic product (GDP)" (Naughton and Segal, 2003), China must reduce its reliance on imported foreign technology by 30% and increase its indigenous innovation capacity by 60% or above, and be among the top five countries in the world for the filing of "indigenous" inventions and the frequency of citations in international journals and science papers. China must also build seven world-class research institutions and universities. China, it is asserted, needs to master core technologies in key technology sectors, including equipment manufacturing and information communication; catch-up with advanced nations in agricultural-related S&T capabilities, development energy, energy conservation and environmental technology; prevent major diseases; develop modern weaponry; and achieve international benchmarks in cutting-edge technologies in sectors such as information technology, biotechnology, materials and aerospace (Gabriele and Haider, 2008:14).

The S&T plans overt focus on "indigenous innovation", responding to the perception among Chinese policy makers, at the time, that foreign technologies dominate in high-tech production networks, relegating China to a costly and low value-added role in the global economy. Indigenous innovation as a strategic policy plan is designed to develop the core technological base and competitive and managerial ability of a firm, industrial sector, regional bloc or nation state (Wang, 2006). The primary goal of indigenous innovation is to explore and develop potential markets through in-house R&D and external knowledge acquisitions.

In the literature, two conflicting perspectives compete for theoretical supremacy. The first perspective emphasises the importance of economies of scale in technological learning, upgrading and innovation. According to Schumpeter (1950) and Chandler (1991), large firms with their abundant resources are better equipped to engage in technological and organisational innovation. In contrast, the external networks perspective points to the importance of dense networking among a large number of competing firms that can create an environment that favours technological innovation and learning (Armin and Thrift, 1992; Pioe and Sable, 1984; Saxenian, 1994 cited in Wang, 2006).

However, China's technological development at this point in the development process did not correspond to either of these models. Instead, it resembled more of a dualist model in which the foreign sector had not established organic relationships with domestic firms, and domestic firms had not built institutional linkages among themselves to encourage collective learning and innovation. The result has been that linkages among both large and smaller Chinese firms remained weak. Furthermore, the foreign sector did not develop production-network relationships with local firms either. Because of China's open-door policy and local government's generous provision of tax incentives, foreign IT firms were funnelled into the high-tech experimental zones of the coastal provinces, starting from Guangdong, via the central coastal areas (Shanghai, Jiangsu, Zhejiang) and to the northern coastal areas

(Beijing, Tianjin, Shandong). In 2002, these three areas accounted for 84% of all IT firms, produced 74.7% of the total value of the information technology industry (IT), and contributed 86.7% of China's total IT export value (National Bureau of Statistics, 2003 cited in Wang, 2006).

It was an attempt to address these issues which led Chinese leaders towards the idea of indigenous development. However, China's indigenous development plan resulted in foreign actors labelling the Chinese government as techno-nationalistic. Techno-nationalists operate from the premise that the Western-dominated international system creates structural inequalities and nations caught on the periphery of the capitalist system are condemned to stay there (Fewsmith, 2007:1). Techno-nationalist policies are designed to create independent domestic capabilities in core or critical technologies. To active techno-nationalists, the construction of domestic institutions designed to coordinate and diffuse these capabilities across sectors and assist producers and users of the technology are key policy goals (Naughton, 1991:4). They argue that a new type of economic war is being waged that involves the standards, rules and protocols which constitute today's technological systems (Suttmeier and Yao, 2004:167). Chinese S&T policymakers in response to this socio-technological environment sought to develop a new developmental agenda that operated from the assumption that "competitive success flows to the company that manages to establish proprietary architectural control over a broad sphere of development" (Suttmeier and Yao, 2004:167).

Standards development as competitive strategy

In the contemporary global economy, international standards have become one of the most important non-tariff barriers to trade, especially national product standards that specify design or performance characteristics of manufactured goods. A technical standard is essentially a patent ratified by government organisations, international organisations or industrial associations or a de facto specification of an industry through market exchanges (Suttmeier and Yao, 2004 cited in Bell and Feng, 2007:54). Divergent national standards often inhibit trade, whereas regional and international standards increasingly serve as instruments of trade liberalisation. Consequently, the setting of international standards – seemingly technical and apolitical – is rapidly becoming an issue of economic and political salience.

The process of standard development can occur either by edict or a process of formal consensus. Standards may be developed at a national or international level. International standards are the product of agreements between member bodies of the International Organization for Standardization (ISO). Essentially, international standards are developed by the ISO technical committees (TC) and subcommittees (SC) (ISO, 2013). For example, in order to ensure the safety and integrity of offshore wind turbines and to address the quickening pace of the development of offshore wind farms in Europe, the need for an international standard for offshore wind turbines was viewed as increasingly important. In order to develop the appropriate international standards the IEC TC88 WG3 was set up in 2000. This brought together international expert knowledge from the wind power and offshore engineering industries and the standard, IEC 61400–3, was developed and realised (Quarton, 2005).

Notwithstanding the growing economic and legal significance of international standards, standards setting has until very recently received scant attention from scholars and non-scholars alike. Standards seemed invisible to all but a few experts in engineering and related fields. In addition, with the exception of some economists and legal scholars, social scientists deemed standardisation unworthy of their attention – the topic seemed hopelessly technical and dry. But is it? The answer is an emphatic no. The study of international standardisation raises the kinds of questions familiar to students of international relations, including: Who sets international rules? Do international standards benefit all or are there winners and losers, either in relative or absolute terms? What is the role of power and institutions in international disputes or bargains over standards? What defines power and how does it operate? (Maltti and Buthe, 2003).

The use of standards to gain market share and block competition has been termed techno-nationalism. Although no standard definition exists, in common usage, techno-nationalism refers to such public policies that "target" the strategic (usually high-tech) industries and give them various governmental support: government procurements, import restrictions, export subsidies, R&D subsidies, R&D tax credits, controls on inward FDI, protection of intellectual properties, government-funded R&D projects, and others. All support is given only to domestic firms (i.e., firms owned by its own citizens), for their aim is to strengthen the competitiveness of domestic industries against foreign rivals in a growing world market (Heires, 2008).

It is important to acknowledge here that the structural politics and conflicts that define the international standardisation process have a significant impact on developing nations. This is because the capabilities to participate in and influence the ISO's work are unevenly distributed. Furthermore, not every standard developed at ISO is equally relevant for all members. However, most standards have a significant impact on the future of technological development. Many countries therefore have a stake in international standards, but developing countries, especially, do not have much say in their development. Without their input, standards, which can also be a means of knowledge and technology transfer, might prove unsuitable to their particular needs. The centralisation of decision making on strategic issues in ISO's leadership and the unequal representation of interests in the technical work of the standardisation committees is therefore inherently problematic (Heires, 2008).

In China, central and local government officials have sought to establish key measures designed to support the creation of domestic IP, including funding for the development of brand names, technical assistance in the preparation of patent applications and the identification of unpatented foreign technologies for exploitation by Chinese companies, and funding to defend patent infringement cases brought against Chinese companies. In the area of technical standards, China has adopted a policy of fostering the development of product standards for a range of electronics products including cellular telephones, digital televisions, integrated circuits, wireless devices and video discs. This standards-setting agenda was constructed in order to decrease China's high dependence on foreign technologies and increase the prominence of standards that rely on domestically controlled IP, in order to increase royalty payments to domestic IP owners. Standards-setting initiatives have been supported by the Ministry of Science and Technology's (MOST) "863" high-technology R&D funding program; at least 29 standards have resulted from this support. One of the highest profile efforts has been the central government's support for TD-SCDMA – a third-generation (3G) cellular telephony standard jointly developed by Chinese and foreign companies – with R&D funding and preferential financing for domestic firms (Lee and Sangjo, 2008).

However, TNCs asserted that Chinese standards-setting processes were defined by a techno-nationalist thrust, in that they are not opaque, not centralised and are closed to foreign participation. Moreover, what I want to emphasise here is that techno-nationalist policies, are becoming an increasing outdated technological development strategy that is highly unlikely to succeed in an increasing globalised socio-technological environment: This is because it is becoming increasingly difficult to create and commercialise high-technology assets without engaging in collaborative behaviour designed to facilitate the co-development and ownership of technological assets and intellectual property.

WAPI

China's first attempt at developing a techno-nationalist strategy was framed around efforts by Beijing to set a mandatory encryption standard for its own wireless mobile security protocol, WAPI. This attempt at standards setting, it has been argued, represented: "a case where China's mercantilist bureaucrats are keen to force foreign interlocutors to work through 'channels of global transaction' that would be controlled and monitored by bureaucrats intent on seeking rents and steering their own version of a developmental path" (Zweig, 2002 cited in Bell and Feng, 2007:54).

At this point in time, the dominant standard that was internationally recognised was the 802.11 Wi-Fi standard that had been developed by the Institute of Electrical and Electronics Engineers (IEEE) (Lee and Sangjo, 2008:665). However, it was widely acknowledged that Wi-Fi possessed security problems. Utilising the security problems associated with Wi-Fi, Chinese policy attempted to set up WAPI as a standard for the domestic market (Lee and Sangjo, 2008:665). Moreover, WAPI did not comply with chips based on Wi-Fi and was to be used as a mandatory standard for both domestically produced and imported wireless networking equipment to be sold in China after May 2004. The technology was proprietorial and had to be incorporated into products in cooperation with one of 24 Chinese companies that had been authorised for this purpose (Linden, 2004:18). A number of these companies were potential competitors with foreign firms, and the terms of trade set by the Chinese meant that foreign manufacturers needed to disclose sensitive information to their Chinese partners. As a consequence, foreign firms that wished to participate in the market would have to build local factories and maintain separate production lines for China and the rest of the world (Lee and Sangjo, 2008:666).

Resistance to the standard was fierce (Bell and Feng, 2007). Foreign businesses were concerned that China may attempt to have the standard adopted globally. The Chinese wireless local area network (WLAN) market is a significant market for foreign investors due to its sheer size and potential for growth. Research conducted by the firm, IDC, indicated that in 2002 China accounted for just US$17.2 million of a US$2.2 billion global WLAN market. Hence, the potential growth trajectory was substantial. The Wi-Fi Alliance further estimated that the growth of the

Chinese market, between the period 2003–07, was US$50 million to US$500 million (Mannion and Clendenin, 2003 cited in Bell and Feng, 2007:55).

Within this period, the Chinese government also sought to integrate its standards with the development of indigenous R&D networks. Its policy directives emphasised the need for commercial development and attempted to involve high-technology start-up companies with direct linkages to government research institutes and universities. However, whilst it is clear that this new fusion of policy with research was intended to operate within the boundaries of China's WTO commitments, specifically the provisions of the Agreement on Technical Barriers to Trade (TBT), the international community was becoming increasingly concerned and questioned the motives of China's non-consultative and exclusionary methods and its intention to try and coerce foreign firms to adopt WAPI (Kennedy, 2006; Suttmeier and Yao, 2004:167).

It was these concerns that served to unite foreign governments and firms in a campaign to address the potential rise of techno-nationalism in China and over the implications relating to China's future economic openness and the protection of foreign intellectual property rights (Kennedy, 2006; Suttmeier and Yao, 2004:167). Intel refused to meet the Chinese deadline for the standard adoption and emphasised the fact that the technology behind the WAPI standard was at least a generation behind current technologies. In an attempt to halt implementation of the standard and address China's growing techno-nationalistic orientation, foreign companies, especially US chipmakers, asked for intervention from their governments, and the dispute developed to involve the US government. The Chinese government, facing mounting international pressure, withdrew the standard for consideration in 2004 (Suttmeier and Yao, 2004:167; Bell and Feng, 2007).

What needs to be emphasised here is that in the case of WAPI, the researchers and holders of the IP rights to this technology came primarily from a small Xi'an-based company named IWNCOMM (*jietong*), and not one of the champions of corporate China nor few of the heavyweight domestic companies in the manufacturing and service provision sectors expressed enthusiasm for this standard. Hence, it is important to note here that in the case of WAPI only a "narrow coalition" supported it. Other actors, such as domestic technology developers, IT manufacturers and telecommunication operators, were not unified; and there existed an extremely well-organised campaign by transnational capital and foreign governments led by Intel to defeat the Chinese government's attempt to impose WAPI (Kennedy, 2006; Zhao, 2010).

Bell and Feng (2007) argue that the failure of WAPI is a direct result of China's deepening reliance on foreign investment, technology flows and export markets, as well as its increased engagement with international regimes and the multilateral rules and constraints on developmental behaviour that this entails. As a result, it is conjectured, China's institutional and structural location within the world economy has fundamentally altered, "creating a structural dependence on foreign interlocutors" (Bell and Feng, 2007).

As a direct consequence of this deepening structural interdependence and the need to participate in the global knowledge economy and its emergent production structures, supply chains, innovation chains, intellectual property regimes and collaborative networks, it has become necessary for the Chinese government to undergo a profound shift in developmental focus. Specifically, it has begun to move beyond the construction of an economic development strategy that is overtly techno-nationalistic in focus. This is because China cannot afford to miss this opportunity to embed itself in the global knowledge economy if it wishes to successfully upgrade its current status as a low-cost producer of simple manufactured products and become a technologically sophisticated innovative nation.

It is here, in this new developmental context, that the twenty-first century's post-industrial architecture is being forged. The development of system-shaping capabilities is, hence, an important prerequisite for successful high-technology development in the contemporary socio-technological system. By developing these capabilities, firms have the best chance of being involved in shaping the architecture of the technological ecosystems around them and of devising key strategies in relation to how they and other systems actors will be organised. Indeed, research conducted by Santos and Eisenhardt (2006) highlighted how even small, start-up ventures can acquire a key position in the industry architecture by influencing the structure of their sector in ways that would eventually fit their own capabilities. Thus, it can be argued that by strategically endeavouring to manage or influence a sector's architecture, a firm can capture a disproportionate amount of the benefits created by an innovation, especially because innovations often require (or justify or legitimise) the creation of a new architecture. This is because opportunities for altering or constructing industry architecture emerges in new sectors, for new technologies, or in areas where a substantial technological, institutional or demand discontinuity exists that requires the reorganisation of production structures, the establishment of relational ties, network effects and legitimation (Dicken et al., 2001). Furthermore, innovation has become inherently more expensive and risky than in the past, with a high concentration of advanced R&D spending by countries and enterprises. Conversely, there is greater inter-firm and cross-national collaboration and networking in the area of innovation (Lall, 2003:12).

Emerging technologies and innovative products and organisational structures are developed in global knowledge and innovation networks. Hence, as the complexity and unpredictability of these technologies increases, and firms endeavour to move beyond the monopolisation of intellectual property rights and develop new strategic action plans structured around the need to formulate organisational and network legitimacy, the role of the state in socio-technological development is being fundamentally altered. From this frame of reference, it is essential that developing nations such as China engage in networking, technology coalition building, political lobbying, and ecosystem alliances in order to transcend its contemporary low-technology base and embed itself in the contemporary global economy in a way that facilitates sustained long-term high-technology development and economic growth (Low and Johnston, 2007).

The key point here is that network capabilities are a key characteristic in the development of relational ties and can lead to returns of economic, social, technical and political capital through the appropriate identification of network opportunities and constraints. Hence, in order to gain network legitimacy in the global economy, the Chinese government has had to undergo a continuous process of policy readjustment and experimentation in order to conform to specific sets of multilateral rules and behavioural norms and technological opportunities – especially in the wake of the WAPI failure.

As the WAPI case highlighted, with its loss of network legitimacy, without the presence of relational legitimacy it is highly unlikely that foreign firms will engage in network activity or disclose proprietorial knowledge on emerging technology to Chinese firms. Committed relational legitimacy is a new form of collaborative activity that provides the Chinese government with a strategic tool whereby technology is not just transferred in a shallow fashion at the level of low-cost manufactured goods, but also is embedded firmly in both the Chinese and the global economy simultaneously, as new forms of long-term relationships and collaborative alliances are formed (Low and Johnston, 2006).

Conversely, as this book will argue, it has become necessarily for the Chinese government to develop a third strategic development plan that entails both the development and strengthening of national innovation policy directives and institutional infrastructures as well as a comprehensive attempt to build an open, market-based framework that considers not just the possibility but indeed the necessity of international cooperation (Yang and Kuo, 2008) and value appropriation. Indeed, a third-generation innovation policy that transcends traditional linear interactive models is beginning to emerge. This policy paradigm is framed around the need to pay theoretical and practical attention to the need for flexible institutional adaptation in the area of science, technology and innovation (STI) policy, the facilitation of relational and network ties, as well as the need to coordinate innovation policy components across ministerial boundaries and hence redefine innovation policy in a horizontal fashion (OECD, 2005). As the OECD conjectures in a report entitled, "Governance of Innovation Systems: Synbook Report", policy directives need to be framed in a flexible and integrative fashion: "While horizontal coherence ensures a strategic, integrated focus on innovation across boundaries and may be supported by cross-sectoral analysis and coordinated reporting systems, vertical coherence ensures follow-up of sector ministerial action plans. Comprehensive innovation policy has much to gain from organising information and learning systems that help policy makers develop an integrated focus on innovation" (OECD, 2005).

The state, therefore, needs to think about industrial development in terms of integrative and flexible capability building, precisely because it has (1) little direct control over the action of individual agents such as firms, research and knowledge zones and international technological and economic organisations such as standards-setting committees, and (2) only a limited understanding of the dynamic nature of global the market. Furthermore, in the specific case of rapid-innovation-based industries, markets and products are not well defined, and hence plans for detailed state intervention are of limited utility. Thus, for a state to have a positive influence on the development of rapid-innovation-based industries, it needs to develop methods other than long-term planning, imitation and pure technology transfer via FDI inflows.

Conclusion

Until recently, the Chinese approach to industrial and technological development has been structured around the attraction of FDI to help leapfrog its economy. It is a policy approach that has clearly failed to deliver the technology and knowledge flows expected. This is because in this period the global economy was defined by coordinated attempts by advanced capitalist firms and governments to promote and regulate global production networks via the construction of international agencies and trade agreements and intranational policies and legislation that operated to assist developed nations' indigenous economic actors to execute transactions that ensure their competitive advantage over foreign counterparts.

Furthermore, the emergence of global innovation networks has rendered S&T policy directives that are overtly national in focus obsolete. From this theoretical perspective, collaborative R&D alliances, joint ventures and the development of network linkages frame the strategic objectives of successful high-technology companies. Strategic policy in this networked environment is primarily conceptualised as a portfolio of links whereby dynamic positioning into large networks is critical to competitive advantage (Gay, 2008:63).

In a contemporary context, the implications of new institutionalised forms of cooperation, organisation and competition that transcend national borders are under-theorised. They are also sectoral specific. This is because innovation and technological change are defined by differing characteristics and follow different paths depending on the sector in which they occur. Indeed, how to strategically embed its firms in these global production and innovation networks and develop long-term strategic alliances and legitimate governance mechanisms at a global level that will allow the government of China to exert a degree of collaborative control over these networks, it can be argued, is one of the most perplexing challenges facing Chinese policymakers today.

3

Complex Global Technological Systems and the Chinese State: From National Indigenous Innovation to Globalised Adaptive Ecology

Rapidly accelerating technological and organisational change at the systems level is fundamentally reshaping the immediate and future socio-technological environment and facilitating new forms of geopolitical interaction, competitive advantage and systematic constraints. Indeed, a new techno-economic paradigm is emerging, and it is these new technologies that are the site for the spatial, relational and organisational transformation of the global economic system. What is so interesting and what comprises this paradigm shift is not just the ways in which these technologies will revolutionise the way human societies and economic systems interact, but the way these technologies co-evolve at a global level, and hence transcend the territorial-based behaviour and economic appropriation models of both the individual firm and the nation state.

Indeed, the emergence of global innovation networks and the need for continual innovation alerts us to the fact that science, technology and innovation policies can no longer be designed solely in a national context. "Go-it-alone" and "indigenous development agendas" with a techno-nationalistic thrust simply do not work in these complex, fast-paced, networked environments. The rapid pace of technological change and the increasing organisational and relational complexity of doing business in these industries make it impossible for a single company – or even a single nation – to compete without technological, financial and organisational collaboration. However, whilst it must be conceded that the ability of national governments to implement indigenous technology agendas is highly constrained by the evolving socio-technological regime, it is important to note that for emerging nations, like China, who were effectively locked out of the previous round of economic and technological development – due to the privatisation and monopolisation of knowledge, the spatial re-organisation of the global economy and advent of more open collaborative models of technological development – provides a window of opportunity for the Chinese nation to redefine its comparative advantage and embed itself strategically in key, emerging, high-technology ecosystems.

Drawing on examples from the global wireless communication sector, the central goal of this chapter is to develop a multilevel, multivariate, conceptual and empirical framework capable of advancing contemporary understandings pertaining to the emergence of the global knowledge economy and the key structural and relational opportunities and constraints that advanced technical systems exert on both firm-level actors and the state. The literature on the evolution of complex socio-technological systems and its relationship to the market and the state is voluminous and theoretically diverse. Conversely, this chapter is not designed to review the entire literature on socio-technological systems and its relationship to the firm, the state and the market, but rather to explore the themes that are important to the development of my central theoretical research questions. Specifically, I am interested in developing a framework to understand how key "ecological" or "systems integration" technical pressures such as the co-production of technology, the occurrence of networked innovation and the increasing use of collaborative alliances for market and product development are impacting on and shaping the socio-technological policy directives of the government of China and the strategic behaviour of key Chinese high-technology firms.

It is the intention of this chapter to argue that technological change and global systems integration via the development and proliferation of scientific and organisational innovation networks, the occurrence of critical ecosystem dependencies and the emergence of new forms of collaborative architecture are facilitating a shift towards a more alliance-based model of capitalist organisation. In this chapter, the concept of alliance capitalism, which currently is overtly focused on the internationalisation of the firm, will be extended to incorporate the notion of reciprocal interdependence whereby firms from differing global locations are encouraged by a range of strong incentives to form trans-territorial technological and market-building alliances. These incentives include an enhanced ability to address the existence of critical ecosystem dependencies, quicker paths to product commercialisation, reduced R&D costs via access to external R&D and co-development, and the ability to engage in the construction of new technological ecosystems and markets and their associated technological and organisational requirements.

Obviously, the emergence of the collaborative economy and alliance capitalism as a key form of economic organisation has fundamental implications for the technological, economic, political agency and strategic action of all actors in the contemporary socio-technological system. Firstly, strategic agency has itself become dispersed and reliant on the capacity of actors to develop and maintain relational legitimacy and ties with diverse sets of stakeholders. Network legitimacy refers to the ability of firms and organisations to "successfully allocate their resources, perform activities and market their competencies to key network stakeholders, by conforming to regulative processes, institutional norms and cognitive meanings within the network environment" (Low, 2010). Secondly, the co-evolution and development of technological assets and intellectual property in a global environment means that a number of new collaborative structural dependencies are beginning to emerge. However, this does not mean that the socio-technological system as an overarching structure shapes strategic agency in a highly determinate fashion. To the contrary, it is the intention of this book to highlight how key ideas and the actions of strategic system actors can shape the socio-technological system itself. That is, "agency" and "strategy" can shape "structure". This ability to strategically shape the socio-technological environment requires an awareness of technological ecosystems and their associated structural features, the emergence and creation of new markets and organisation methods, and the ability to participle and coordinate activity in multiple technological arenas simultaneously. Moreover, a number of identifiable roles and associated capabilities have recently emerged that indicate that early attempts to shape the socio-technological system are already emerging and that these strategic shapers will play a key role in these technological systems' innovation platform architecture development, coordination processes and value apportion processes. These system roles and capabilities include: focal and network organisers, system integrators and strategic technological niche development.

Conversely, in order to effectively operate in this environment the government of China is presented with a set of strong incentives to engender a strategic shift away from ideas of indigenous technological development and national sectoral pillars to a socio-technological development strategy that is informed by the concept of "globalised adaptive ecology", which is a strategic policy response that combines state agency and developmental planning with the need to gather, co-create and appropriate technological knowledge and assets from multiple states and technological knowledge zones. It is also a highly malleable policy framework that is designed to be continuously redefined as the architecture of the system itself is subject to dynamic change.

The idea of "globalised adaptive ecology" and "structural socio-technological system shaping" provides us with a powerful medium from which to examine the adaptive behaviour of the Chinese government and Chinese firms and their ability to operate and compete in multiple technological domains and system levels. The key concepts that will be incorporated into this framework are: co-evolution, alliance capitalism, self-organisation, complex adaptive ecology, dynamic structural change, knowledge coordination, experimental trial-and-error learning, strategic system shaping, actor agency and trans-territorial multilevel governance.

In order to evaluate the increasing occurrence of collaborative alliance-based behaviour and the co-development of high-technology products, this book will empirically examine the alliance-based behaviour and co-development product activity and network connections of three key Chinese high-technology firms in Chapter 6: China Mobile, Huawei and ZTE. It is important to note here that the emergence of critical ecosystem dependencies and the shift to a more alliance-based economy is applicable to a number of high-technology sectors. However, the focus of this book is on the global wireless communication sector only.

Ecological and socio-technological systems integration

In the twenty-first century, the emergence of large-scale global socio-technological systems has profoundly modified how both states and firms interact, and how states govern technological development. In this new socio-technological economic system, future economic growth is located in emerging high-technology sectors such as biotechnology, genomics, cloning, nanotechnology, the green economy, microelectronics and information communication technology. These technologies are themselves highly ubiquitous, have the power to integrate and combine in new forms and possess the power to disrupt existing economic models and societal orders on a faster and larger scale than ever before in human history (Khanna and Khanna, 2010).

Certainly, the creation, dissemination and appropriation of knowledge has become a major engine of economic expansion in the twenty-first century global economy. For instance, global R&D expenditures over the past decade have grown faster than GDP – an indication of widespread efforts to make economies more knowledge and technology intensive. For example, global total R&D has risen from an estimated $522 billion in 1996 to approximately $1.4 trillion in 2012 with the rate of growth slowing only in the 2008–09 recession year. Although the specific data is comprised of approximate estimates, the sustained and strong upward trend highlights the rapidly growing global focus on innovation through R&D. Most of the global funding growth is being driven by Asian economies, primarily, who were expected to increase their R&D expenditure by nearly 9% in 2012. In stark contrast, in the same year, European R&D was expected to grow by approximately 3.5%, whilst North American R&D growth was forecasted at 2.8%, albeit off a higher base (OECD, 2012).

However, it is important to note that the globalisation of R&D is not the only defining characteristic of the new innovation landscape. Innovation not based on R&D, including non-technological innovation, is increasingly perceived as an important contributor to economic growth and development. Specifically, innovation surveys find that a large share of innovative firms do not conduct any formal R&D at all. For example, almost half of innovative firms in Europe do not carry out R&D in-house. Instead, they are focused on the development of new organisation methods, forms of network coordination, collaborative connection generation and knowledge sharing. Indeed, recent research suggests that process and organisational innovation are becoming prominent drivers of improved firm performance globally (OECD, 2012). Hence, comparative advantage is located in the adaptive, organisational and relational capacities of the firm and the state.

For example, Procter and Gamble (P&G) have recently moved from a closed in-house innovation system to an open innovation system structured around the idea of "Concept and Development" (C&D). By utilising this more open innovation platform, P&G aims to become the open innovation partner of choice (Joia, 2009). To achieve this goal, P&G actively searches for external collaborators to develop long-term relationships and joint development opportunities. These external sources of innovation assert P&G is for everything – from concepts to how to go to market: "Finding them requires a robust pro-search effort across the globe, looking at every region, every country, every company, every university, and identifying what expertise exists" (Joia, 2009). When the process began, 10–15% of the innovations at P&G included ideas from external sources; and today this has increased to 50%, resulting in the company achieving 6% organic growth in an industry which is growing at 2–3% annually (Lafley, 2008 cited in Ruiz, 2009).

Furthermore, product modularity and architectural innovation have increasingly reframed the technological development process in recent years. Product modularity refers to how components are integrated into products via sub-elements, sub-assemblies, sub-systems or "modules" that independently perform distinctive functions. "Module knowledge (also called component knowledge) focuses on these modules (components) themselves as opposed to the linkages between components" (Narayanan and O'Conner, 2010).

Architectural innovation refers to the linkages that occur between the components of established products in new ways, whilst the core design elements remain untouched (Henderson and Clark, 1990). For example, the Sony Walkman can be cited as an example of an architectural innovation, where miniaturisation of radio technology facilitated portability, hence significantly altering the way in which music was listened to. Indeed, the Walkman fuelled an entire industry of portable music players. Conversely, it has been conjectured that "modular innovation involves the introduction of new technology to specific modules of a product that displaces the core design concepts while leaving the established linkages between components relatively untouched" (Narayanan and O'Conner, 2010).

Indeed, the emergence of new sectors, business models and structural changes in the socio-technological system coupled with the rapidly accelerating pace of technological change at multiple levels indicates that technology is a major agent of both system change and structuring in the contemporary socio-technological system. Specifically, the concepts of "convergence" and "interoperability" are key structural change drivers that play a fundamental role in shaping the behaviour of system actors and thus need to be accounted for in theoretical and empirical modelling.

Conversely, in order to participate in the global networks and sectoral systems that make up the contemporary socio-technological system, actors need to develop technological innovations that possess the ability to converge or interact with other firms' products and evolving technological platforms. Furthermore, the ability to ensure "convergence" and "interoperability" is highly dependent on the capacity of firms to access globally dispersed scientific and technical knowledge sources and generate long-term networked relational ties with other system actors. As a direct consequence, innovation occurs within an ecosystem of interdependent innovations (Adner and Kapoor, 2010). Conversely, economic innovation and technological development in the contemporary global knowledge economy can be said to occur within a context of dynamic uncertainty and is defined by the possibility of multiple outcomes and system forms. At the present time, this multilevel technological environment is comprised of a number of loosely configured technology zones and a set of more formalised institutional precepts. These include innovation ecologies, socio-technological regimes and emergent technological niches (Barry, 2006). Within this dynamic technological environment, the firm and the state are both rooted in a dense array of deterritorialised or partially territorialised alliances with public- and private-sector actors (Haberly, 2011). This means that, the development of strategic socio-technological innovation strategies and associated capabilities must be framed by an understanding of how technology facilitates and shapes the emergence of socio-technological systems and how exactly these large-scale global technological and knowledge systems are being politically, economically, socially and scientifically shaped and organised at the national, global and sub-national level. Thus, in order to comprehend exactly how the emerging collaborative economy and the contemporary global socio-technological system affect the behaviour and developmental capabilities of their members we need to understand how the they function and how the key variables encapsulate their evolution, development and governance. This requires an understanding of exactly how complex adaptive systems contribute to technological evolution and socio-technological systems development.

Complex adaptive systems and technological evolution

The theory of socio-technological complexity is derived from evolutionary economic theory. Primarily, complexity theory and its numerous theoretical derivatives endeavour to offer a paradigm cable of explaining the structural and dynamic properties of technological knowledge generation and diffusion as it relates to the emergence of innovative technological knowledge assets, institutional forms and economic development (Antonelli, 2010). Specifically, complexity theory is both systematic and dynamic, and it provides a set of theoretical tools that help us to conceptualise how knowledge production occurs, is diffused and exploited, as well as the emergence and transformation of system architectures for the coordination of knowledge and appropriation through time (Patrucco, 2008).

Socio-technological complexity theory operates from the assertion that the current global socio-technological market system is an example of a complex adaptive system (CAS) that is composed of a decentralised collection of networked actors interacting in various market contexts. Technology systems, as complex adaptive systems, are characterised by their dynamic interdependencies across various scales and are driven by mutual interactions between institutional, ecological, technological and socio-economic domains. Hence, because the contemporary global socio-technological system is defined by the existence of dynamic and highly interdependent developmental processes, no single technological artefact or developmental process can be isolated for theoretical analysis. Thus, it has been asserted that given the fact that technological development and innovation in a contemporary context is so interrelated, technological development is best conceptualised as occurring in an ecological system of co-evolving artefacts (Geels and Schot, 2007). For example, in the global wireless communication sector, each of its three key segments (applications, devices and core networks) is highly dependent upon innovation in the other segments for value creation. For instance, new applications depend upon both advances in device hardware capabilities and advances in the spectral efficiency of the core networks to provide the network capacity to serve those applications (Faulhaber and Farber, 2010).

Another important tenant of CAS theory is the concept of co-evolutionary potential. The central idea of co-evolutionary potential is the inherent need of systems actors to sense and respond to feedback and the ways in which this feedback ultimately spurs mutual and dynamic interactions between the particular sub-systems or evolving elements. Hence, in CAS, strategic technological policy development often resembles a "co-evolutionary dialogue" where a continuous learning process is driven by the mutual and reciprocal interactions among the interlinked sub-systems and agents. Alongside this "dialogue", the ability to form new relations and new emerging properties enhances the chances of adaptive change and social-ecological resilience. This emphasises the inevitable interdependencies between technologies, institutions, values and the emerging systems' relational architecture (Nye, 1992; Vicenti, 1994; Mokyr, 1990; Basalla, 1988; Nelson and Winter, 1982 cited in Rammel, Stagl and Wilfing, 2007).

For example, in The Wide Lens, Adner asserts that the historical focus of companies on managing their execution risk – that is, coming up with innovations that are valued by their customers, and delivering them better and faster than the competition – has resulted in what he terms a "blind spot" (Adner, 2012). This "blind spot" arises because by focusing solely on product development and customer need, critical ecosystem dependencies are neglected. As a direct result, even highly innovative products with a wide customer base fail to reach the market (Adner, 2012). The key problem here is that in many contemporary global knowledge economy sectors, bringing an innovation successfully to the market requires partners that are both able and willing to participate in the development of complementary products and service solutions. Conversely, Adner asserts that contemporary innovators need to be aware of two new kinds of risk: co-innovation risk and adoption chain risk (Adner, 2012). Co-innovation risk asks the question: "To what extent does the success of an innovation depend on the successful commercialisation of other innovations?" What is important to emphasis here is that customer insight and innovative development efforts will not make up for the absence of complementary assets (Adner, 2012). Adoption chain risk considers the question: "Who else needs to adopt my innovation before the end customer has a chance to assess the full value proposition?" The issue is: What are the incentives for partners to participate? The key to managing this risk is to recognise that alignment with the interests of adoption chain partners is as important as delighting end customers (Adner, 2012). Examples of firms that have been disadvantaged by a lack of complementary ecosystem assets include Philips. In the 1980s, Philips made a significant investment in order to develop its high-definition television (HDTV) sets. However, the project failed when other firms failed to develop the high-definition cameras and transmission standards needed to make HDTV work. Philips incurred a US$2.5 billion write down at the time (Adner, 2012).

Another pertinent example is Nokia. In a bid to be first to market with a 3G handset Nokia invested millions in the development of the devices. However, it failed to achieve first-mover status precisely because its ecosystem partners did not complete their innovations in time. By the time the necessary complementary assets and products, such as customised video streaming, location-based services, and automated payment systems, were ready for the market, so was the competition (Adner, 2012).

These examples of ecosystem failure can be contrasted with a number of innovations whereby firms recognised that their success was dependent on the development of a complementary ecosystem. For example, the Digital Cinema Initiative (DCI) is a consortium of movie studios that came together in a unique way to overcome the cost of adopting digital film within the theatre value chain. In essence, they subsidised and shared the cost of capital investment in smaller theatre chains to ensure that digital film would enjoy the broad distribution and availability critical to its growth (Moore, 2012).

It is clear that these theoretical propositions are highly applicable to the global wireless industry, where a firm's value relationships are embedded in a larger flow of activities that can be called the "wireless value ecosystem" (Pagani and Fine, 2008). Such large-flow activities create unique sets of ecosystem dependencies that need to be addressed by all ecosystem actors in order for the system to function effectively. On the upstream side, contractors and electronic manufacturing services (EMS) create and deliver the purchased inputs used in an equipment manufacturer's value relationships. On the downstream side, network operators (initially, subsidiaries of the national postal, telegraph and telephone service (PTTs) or carriers) serve as distribution channels and perform additional activities that affect the buyer. The ultimate basis for any differentiation is the company and its product's role in the buyer's value relationships. As a firm, each company in the wireless industry has its own value relationships; as industry players, all take part in the broader industry value system. Achieving, sustaining or renewing strategic advantage depends not just on an individual firm's value relationship but on its role in the broader value ecosystem. These value systems are not static; they change over time. For example, Apple Inc. is the leader of an ecosystem that crosses at least four major industries: personal computers, consumer electronics, information and communications. The Apple ecosystem encompasses an extended web of suppliers that include Motorola and Sony, and a large number of customers in various market segments (Moore, 2012).

Moreover, in the global wireless communication sector, manufacturers of handsets and carriers must work closely to ensure that both the phones and the networks function in a way that ensures both quality transmission and spectrum efficiency. As carriers develop new protocols to increase network performance, they must work with the device makers who will build the handsets in order to achieve system convergence and interoperability. Consequently, the development of industry standards is increasingly occurring jointly in standards committees where both carriers and device makers can represent their interests in order to enable innovation on both sides of the market to occur. Due to the complex nature of the market, which means that no single actor can provide a service to the customers with an end-to-end solution on its own, there is a need to sustain viable alliances and to create a value network with the right partners (Barnes, 2002; Pigneur, 2000; Sabat, 2002 cited in Pagani and Fine, 2008).

Pagani and Fine (2008) use a five-pronged diagram framed around the content value chain, application value chain, infrastructure value chain, network value chain and device value chain in order to highlight the complexity and co-evolutionary nature of the global wireless value system. They highlight how the value system has co-evolved over time and includes a content value chain, application value chain, infrastructure value chain, network value chain and device value chain. Examples from this co-evolving value chain include: mobile phones, PDAs, smartphones, platform infrastructure services, gaming, messaging, MP3 players, circuit board component manufacturers, network equipment providers, wireless network operators, portal access providers, application developers, voice and data, etc. All are linked by dynamic interaction and a casual looping that operates to define the developmental path of the sector over time.

Conversely, the idea of co-evolutionary potential and adaption provides theoretical insight into the conditions under which agents or clusters of agents change their schemata (routines) and how this in turn changes their fitness functions (selection mechanisms) that lead to systemic co-evolution (Holland and Miller, 1991; Levinthal, 1997; Miller, 1992 cited in Geels and Schot, 2007). From this frame of reference, co-evolution is essentially a state of dynamic equilibrium, in which heterogeneity in adaptation performance among agents constantly reshapes the competitive landscape where the selection mechanisms operate. Leading agents or clusters of agents, which adapt faster or more efficiently, constantly exert a pressure for higher fitness that at times results in the fundamental transformation of the system. Thus, the composition of agents, the distribution of agents' schemata and the connections between agents are subject to constant change and evolutionary pressures.

Structure, agency and globalised adaptive ecology

It can be argued that in this new more collaborative environment, achieving first-mover status is not the only way to move up the technology ladder and acquire market share. Instead, ecosystem shaping, co-development and system embeddedness are new strategic models of behaviour that can be utilised to facilitate the successful adoption of a firm's products in the contemporary market system.

Until recently the study of global innovation and production networks (GIPNs) (Gereffi and Korzeniewicz, 1994, 1996) or global production networks (GPNs) (Borrus et al., 2000; Ernst and Kim, 2002; Yusuf et al.,

2004) was primarily structured around the idea of organisational space, and assumed a hierarchical governance structure dominated by either commodity-chain drivers or network flagships (Lee and Saxenian, 2007). From this theoretical framework, system drivers or flagship firms occupy the highest hierarchical layer and assert control over system participants at lower levels via their control over critical assets such as de facto standards, intellectual property rights, brand names, marketing channels and technology production. It is this hierarchy that serves to divide the hierarchy from lower levels. On the one hand, although the high tiers continually shed capabilities, they still maintain their dominance by continuing to control critical assets. On the other hand, although there is an upgrading path providing for lower tiers, this path has a glass ceiling because the power asymmetry works to the disadvantage of lower tiers in the head-on competition with higher tiers (Lee and Saxenian, 2007).

However, this system of hierarchal control via intellectual property and organisational control mechanisms is becoming an increasingly outdated framework for conceptualising global systems innovation in high-technology sectors. This is because in the emerging collaborative economy and its associated networks and sectors, the network does not select for intellectual property ownership per se, but rather the ability of system actors to engage in collaborative behaviour in order to facilitate the co-production, commercialisation and organisation of innovative knowledge assets. This has resulted in the emergence of a new form of network organisation, the Collaborative Global Innovation Network (CGIN). This new model of global innovation has created new sets of system opportunities for economic and technological actors.

The advent of CGINs and the associated critical ecosystem dependencies that permeate these contemporary technological systems is an important example of structural change in the contemporary global economy. The ability of innovating actors such as the Chinese state to engage in transformative high-technological development is hence mediated by these structural features and the way in which they frame contemporary socio-technological development processes. However, what is important to note here is that the structures and institutional frameworks that define contemporary CGINs are a direct product of actors' past decision-making processes and selection mechanisms. Thus, structures are always mediated by actualised agents (Bell and Hindmoor, 2014). From this theoretical perspective, structures can both constrain and empower agents. In order to expand the range of contexts for analysis in which agents interact, Bell and Feng (2013) have developed an "agents in context" approach, which expands the range of contextual variables that need to be examined when trying to locate the source of institutional and structural agency that strategic actor-agents employ when trying to affect socio-technological change. These include: the power of the state, the structural power of businesses, material incentives or disincentives and strategically selective terrains (Bell and Hindmoor, 2014).

In the contemporary socio-technological system, the ability of the Chinese state as a developmental actor to mediate these structural opportunities and constraints in a way that is commensurate with its high-technology developmental goals will be greatly enhanced by its ability to either build new technological ecosystems or infiltrate existing networks and technological ecosystems and play a role in shaping the interests of key system actors and contributing organisational, financial and technological resources for the further successful development of the whole ecosystem. From this theoretical perspective, agency involves the "propagation of innovations" by questioning and replacing "business as usual" policies and creating alternative practices, thereby challenging the established world views and paths, and attitudinal and behavioural patterns, as well as providing others who think as they do (i.e., followers, early adopters) with a constant motivation for a self-sustaining change (WBGU, 2011:243).

Bell and Hindmoor (2014) assert that agents such as the Chinese state act in a myopic and self-interested manner, producing large structural effects they do not fully comprehend or control. The power in question, they argue, is not material or ideational, but rather is a form of structural or ecosystem power born of myriad agent interactions in a structured context, or "a form of power generated by unplanned, uncontrolled interactions between agents" (Bell and Hindmoor, 2014).

Agency is thus understood in a systemic way to spread its effect and power from one scale or level to the next, depending on opportunities and resources, alliances and circumstances. Agency for strategic change is conceptualised here as being embedded in and dependent on the evolutionary processes of selection, variation and retention – starting from within niches and diffusing across scales. And once again, these evolutionary approaches to social innovation rely on the market as the central selection mechanism, or at least as the central metaphor organising their understanding of change and agency (Hausknost and Hass, 2013).

Conversely "globalised adaptive ecology" as a theoretical framework needs to model this increasingly globalised and collaborative economy in a manner that captures the way in which the coordination and governance of scientific and technical knowledge is goal directed by system actors in order to ensure system interoperability as well as how the socio-technological system is characterised by a high degree of complex, self-organising and adaptive behaviour at a global level.

One key strategy used by innovating actors to shape the technological direction, institutional infrastructure and organisation processes of the emerging technological system is to actively construct a technological niche or ecosystem. Technological niches provide actors with an innovation launching platform that helps them to insert themselves into the socio-technological ecology and develop strategic action plans before regimes or more institutionalised forms of behaviour are formed. It is here, in these clusters and networks of experimental technological and organisational development, that new institutional forms and organisational behaviour begin to develop.

Strategic niches and technological development

The niche is a central construct that describes the position of an organism or species in an ecosystem. A similar concept has been applied extensively in organisational ecology to describe the position of an organisation or organisational form in a socio-technological population or community, respectively. It is argued that the niche of an organisation (or an organisational form, for that matter) is the locus of competition, legitimation and selection (Hannan, Carroll and Pólos, 2003).

As I have already previously established, achieving first-mover status is not the only way to move up the technology ladder and acquire market share in the contemporary socio-technological system. Latecomer firms that are engaged in technological upscaling can also engage in a number of alternative strategies, such as ecosystem shaping, co-development and ecosystem embeddedness.

In fact, it is becoming increasingly clear that in the contemporary knowledge economy, which is defined by sets of collaborative ecosystem dependencies, the pursuit of first-mover status may not be best practice for market success and technological value appropriation. Instead, it has been argued, that it is important that firms consider the parameters and dependencies links that will define their specific ecosystems before developing a strategy. From this theoretical perspective, "it only pays to be the first mover if your innovation has very little dependency on the ecosystem. The more complex your innovation becomes and the more it depends on co-innovations and adoption, the less beneficial it is to be a first mover (i.e. the risk of moving first goes up significantly)" (Adner, 2012).

For instance, recent firm examples suggested that for second and subsequent movers it can be more valuable in the short term to build a copycat product that plugs into an existing ecosystem than it is to build a first-to-market innovation. As Adner argues:

> scale matters, and networks and ecosystems give you that scale. To be a great business in today's digital world, it requires spotting all the emergent technological trends on the borders and edges and transforming them into new, scalable market opportunities that build on existing strengths in a unique way. The classic example is Microsoft Windows and the Apple Macintosh computer. Apple had first-mover advantage, delivering the Macintosh in 1984 while a truly-useful version of Windows didn't appear until 1990. How did Microsoft do it? Steve Jobs said Bill Gates stole from him, and in fact he didn't even build a "better" mousetrap. He built one that was good enough. So how did Microsoft win? It won by building a bigger ecosystem, by bringing more people and companies into its orbit. It let others build Windows PCs. It didn't demand that everything under its banner be super-great. It didn't go for the last dollar. Mastery of the ecosystem is the great strength that made Apple the supreme success story of our time. (Adner, 2012)

Furthermore, what the Apple and Microsoft examples highlight is how important addressing critical ecosystem dependencies is for the successful commercialisation of technological products. For example, Apple's iPod was successful after earlier MP3 players had failed, not only because it was simple to use but also because Apple waited until broadband technologies were sufficiently developed in order to support the music data transfers it was reliant on. This created an emergent ecosystem that Apple further enlarged by introducing the iTunes Music Store. This ecosystem was then further enlarged when Apple opened up the Mac-only device to PC users. Hence, it can be seen that in the case of Apple, first-mover outcomes have been very poor for their most inventive products (e.g., Lisa, Newton). However, "when Apple has followed, with incremental innovation (e.g. iMac, iPhone, iPod, iPad), they have achieved much greater success, outperforming both their own first movers, as well as the first movers in each of these categories" (Darner, 2012 cited in Blackenhorn, 2011). As Adner, concludes: "Aligning, enticing, and – occasionally – subsidizing partners are the new ante in the ecosystem game. Amazon and Apple will go down as case studies in alternative strategies for succeeding in ecosystems. Their product-focused rivals will illustrate what it means to be 'stuck in the middle'" (2012).

System integrators and network organisers

A number of identifiable roles and associated capabilities have recently emerged that indicate that early attempts to coordinate and shape the socio-technological system are already emerging and that these strategic shapers will play a key role in the evolution of the technological systems and value appropriation processes. Three main agent types can be identified in the contemporary system as key strategy-shaping actors in this global socio-technological system. These are: systems integrators, focal organisers and network organisers.

The primary goal of systems integrators is to bring the component sub-systems into one system and ensure that the sub-systems function together as a system. Hence, the role of the systems integrator is a more technical role. Systems integration involves integrating existing, often disparate systems. Systems integration is also about adding value to the system, capabilities that are possible because of interactions between sub-systems. In today's connected world, the role of systems integrators is becoming increasingly important because more and more systems are designed to connect, both within the system under construction and to systems that are already deployed. Primarily, systems integrators need to develop and source new technologies and ensure interoperability and convergence can occur for diverse and locationally dispersed participants. This can involve industry alignment, partner selection, pre-competitive market construction and infrastructure development.

Focal organisers are focused on the provision of early funding and network development in order to facilitate what is termed "first to the world" basic research. It is this early research that provides the key platforms for ecosystem building and development. Governments and large multinational firms are often focal organisers due to their ability to invest risk capital and wait for longer periods for capital returns.

Network organisers play a key role in organising system participants as well as in the overall governance of technological niches, ecology and zones. The role of the network organiser is to shape global ecosystems and thereby fundamentally influence the future trajectories of socio-technological industries and markets. Network organisers drive ecosystem collaboration by recruiting members and working to set the direction and timing of sustained collaboration over time. Network organisers seek to ensure that the ecosystem is designed to provide a win-win value proposition for all participants. This is because pursuing strategies that ensure one member's success at the expense of other members does not work in these sorts of collaborative ecosystems – precisely because the successful co-development and adoption of products from the ecosystem are a prerequisite for successful market entry and commercialisation of all system actors.

Thus, a key role of the network organiser is to align the interests of all ecosystem members towards the pursuit of a common goal. Conversely, a network organiser needs to possess the ability and the commitment to create the ecosystem structure, establish fair standards and consistency, and convince potential followers that there is value in it for them. Primarily, this entails that the network organiser is willing to make the initial investment required in order to kick-start the ecosystem. This often requires that the ecosystem leader capture value and collect its rewards in the end, not the beginning. For example, in the case of the e-book, Amazon made certain to keep all of the players aligned and showed a willingness to sacrifice some margin to the publishers in order to ensure that they would feel less threatened and more fully incentivised by the new development, and this has been one of the keys to the Kindle's success (Adner, 2012).

Secondly, it can engage in the development of new organisational forms and methods that are designed to unite and drive technological participants in new ways that challenge exciting technological regimes and zones. This involves the generation of new innovation platforms that conspire to both form and reshape existing institutional models, identities, practices, and relationships, and reduce uncertainty in the socio-technological environment.

For example in Chapters 5 and 6, I will highlight how China Mobile, as a state-owned flagship firm in a highly successful attempt at ecosystem building and shaping, constructed a global collaborative platform for the co-development of time division duplex (TDD) and frequency-division duplexing (FDD) standards and network devices in order to appropriate value and network embeddedness in the emerging 4G market.

It is important to note here that the assets of the shaping entity play a fundamental role in persuading potential participants to invest in the shaping strategy. These include organisational resources, relational legitimacy, intellectual property portfolios, and market access and network embeddedness. In this domain, large established companies and governmental actors have a potential advantage as strategic system shapers. This is because system participants believe that these companies or governmental actors possess the necessary resources to support and sustain the shaping strategy. Consequently, these actors are able to motivate a large number of system participants to make significant investments and take aggressive action to accelerate movement towards a preferred outcome. It also provides the focus and incentives necessary to unleash distributed innovation as thousands of specialised participants experiment to meet shifting and emerging customer needs and business opportunities (Hagel, 2008).

Intellectual property and the collaborative economy

In a more collaborative alliance-driven economy, the privatisation and monopolisation of knowledge as a development strategy, as expounded by earlier Western models of value appropriation and the more recently devised Chinese indigenous innovation strategy, have decreased in utility. In stark contrast, the ability to create and generate relational ties that allow system actors to connect more rapidly and effectively with others to create new knowledge has assumed significant strategic value. That is, stocks of knowledge become progressively less valuable while flows of knowledge – the relationships that can help to generate new knowledge – have significantly increased in value. Conversely, it has been asserted that in the emerging socio-technological system rather than jealously protecting existing stocks of knowledge, institutions and firms need to offer their own knowledge as a way to encourage others to share their knowledge and help to accelerate new knowledge building (Hagel, 2008).

Hence, it can be seen that the capacity for "connectivity", which is associated with the potential of the system to establish relationships and generate interactions with other systems with the objective of increasing their knowledge and technological base, is an important prerequisite for socio-technological systems participation. For the Chinese government, the strategic objective here is to find ways to harness the increasing modularity of this new socio-technological system in order develop the necessary institutional and relational assets that are essential for both the co-creation of intellectual property assets as well as new modes and diversified forms of IP value appropriation and sharing.

This ecological system change towards a more open, collaborative and relationship-based model of intellectual property development and appropriation provides the Chinese government, its key domestic firms and the international trading regime with a distinctively new, more inclusive global innovation system. The strategic challenge here is to attract external innovators who can "co-create value" by identifying new functions and designing novel solutions for a given technological system. In such systems, firms will both give and take IP from the ecosystem (Henkel, Baldwin and Shih, 2012).

The increasingly systematic nature of this collaborative behaviour is seen manifest in the emergence of new "collaborative IP mechanisms", such as IP clearinghouses, exchanges, auctions and brokerages, model agreements and frameworks for IP sharing (OECD, 2011a). For instance, firms increasingly say that they are organising licensing activity and strategic alliances around an IP strategy that seeks to share technologies rather than to use IP solely as a defence mechanism. For a number of firms, this represents a fundamental change in IP management and strategy (OECD, 2011). Moreover, recently a new form of third-party innovation intermediary or "inmediary" – has emerged around the world. NineSigma, InnoCentive, Yet2.com and YourEncore are a few examples. These intermediaries facilitate and broker collaboration across technology markets by providing innovation platforms that link companies with potential problem solvers and facilitate the diffusion of knowledge or technologies (Chesbrough et al., 2011).

The intellectual property and products developed in this new IP system are defined by multiple owners and modes of value appropriation. For example, in order to respond to this increasingly fast-paced and complex socio-technological environment, Nokia has developed a network of nine satellite design studios in various target geographies such as Brazil, China and India. Design teams collaborate with local partners (e.g., Srishti in India) as well as across the global studio network to develop new lines of phones. Another example can be drawn from the recent announcement by Taiwan Semiconductor Manufacturing Company (TSMC) to develop an "Open Innovation Platform" to improve collaboration among customers, electronic design automation (EDA) software vendors and chip architecture IP providers. TSMC hopes to make it easier for customers to access various design tools, foundry data and factory process models to accelerate chip-design-cycle time and improve time to market. Another example, is how computer chip manufacturers, such as Nvidia, provide "reference designs" which include sample hardware implementations and driver software. In this fashion, they split their own IP: keeping the majority proprietorial, they also create "open" modules that include all the designs, electronic files, and test programs that a systems manufacturer might need (Henkel, Baldwin and Shih, 2012).

Furthermore, recently released data filed under the Patent Cooperation Treaty (PCT) indicate that the frequency of co-inventions (i.e., patents with several inventors listed as applicants) is increasing. It is interesting to note that an industry breakdown of joint patents highlights the fact that emerging high-tech sectors contain the most co-owned patents. For example, the chemical industry (at 32.1%) and the information technology sector (at 26.7%) contain the largest percentage of jointly owned patents. This is followed by instrumental pharmaceuticals, including pharabiotenology (11.5%), the automotive industry (5%) and aerospace and defence (3.7%). In stark contrast, sectors such as food and beverages, steel and metals, and oils and gas were not accounted for in any patent co-ownership figures (Hagerdooorn, 2003).

Moreover, in the 4G sector, a number of patent pools have recently been set up to make it easier for companies to license the technology they need to implement a given standard and to prevent costly patent litigation in the future. For example, in 2008, Worldwide Interoperability for Microwave Access (WiMAX), a key 4G technology, alongside Long-Term Evolution (LTE), a 4G wireless communications standard developed by the 3rd Generation Partnership Project (3GPP), brought about a change in how patent licensing is addressed. This involved the creation of a single organisation, the Open Patent Alliance (OPA), which has brought together major vendors including Intel and Cisco Systems. OPA members include Acer, Alcatel-Lucent, Alvarion, Cisco, Clearwire, Huawei Technologies, Intel Corporation, Samsung Electronics, Beceem, GCT Semiconductor, Sequans, and UQ Communications, indicating broad 4G ecosystem support for the pool. A WiMAX patent pool, it is asserted, will help to ensure product differentiation and interoperability at a more predictable cost through a more competitive royalty structure. This, it has been conjectured, is critical to the long-term success and broad adoption of WiMAX, by enabling a wider array of devices such as smartphones, consumer electronics and PC industries to quickly integrate the latest wireless broadband technologies into their products. Furthermore, as of October 2012, a new 4G LTE patent pool has been formed. The patent pool managed by Dolby Laboratories, Inc., a subsidiary of Via Licensing Corp., includes: AT&T, Clearwire, DTVG Licensing (DIRECTV), HP, KDDI Corporation, NTT DOCOMO, SK Telecom, Telecom Italia, Telefónica and ZTE Corporation (Reddy, 2011). Moreover, Via President Roger Ross says his company is talking to another 26 patent holders, including "a couple of chip companies", and he expects the number of companies in the patent pool to grow in the near future. All 26 of the companies involved hold at least one standard essential patent, meaning they hold a patent that must be licensed in order to make products that meet the requirements of standards bodies

(DeGrasse, 2012). Clearly, these patent pools represent the emergence of a more open, collaborative or mixed (combination of open and closed) approach to intellectual property ownership and a fundamental change in the way the technology will be developed and deployed.

For example, in the global wireless communication sector, the transition from 1G to 4G has been defined by a shift from proprietorial technologies and patents to increasing openness and standardisation. In the past century, the wireless value system has moved from domestication towards globalisation in a destabilised non-linear evolution via three stages: monopoly, transition and competition. The monopoly era began in the late nineteenth century and was dominant well into the 1980s. It includes the pre-cellular phase and the 1G era, in which a single-network operator, typically the national PTT, controlled the wireless value system in most developed-country markets. Via the possession of a natural monopoly, these national PTTs controlled the system, from the supply chain and firm activities, to the channel and buyer-value relationships, to contractors and manufacturing, to marketing and customer service (Steinbock, 2003:208). When the transition to 2G began, wireless revenues were still marginal and ancillary to most telecom giants, who focused on corporate markets and high-end consumer markets. As Steinbock (2003) articulates:

> Until the 1980s and 1990s, the interplay between the dominant network operator and its supply relationship with one or more equipment manufacturers served as the prime catalyst of change and winning value propositions. The operators dominated the relationship through the monopoly stage, whereas the competitive vendors captured the bargaining power with the great transition. In good times, both parties have seen the relationship as symbiotic and complementary; in bad times, as parasitic and adverse. Until the 1980s and 1990s, this "Red Queen effect" caused a substantial amount of evolution in the operators and vendors, without necessarily effecting long-term changes in either. Both sides seemed to be evolving in place, like hosts and parasites. With new technologies came new policies and strategies, which subjected the entire value system to transformational industry forces. Just like organisms in biological ecosystems, many important business adaptations originate during times of intense co-evolutionary pressure. Today, the industry provides the growth core for multiple global players in several stages of the value system, covering both corporate and mass consumer markets worldwide. The system is far more complicated. It includes contractors (components

contract and EMSs), subcontractors, equipment manufacturers, platforms, enablers, content aggregators, retailers, and network operators. Moreover, the system is rapidly specializing and globalizing (on globalization, see Bartlett and Ghoshal, 1998). Because of accelerating financial stakes, the control and coordination of the system has become an issue of international competitiveness, particularly in the most developed markets of Western Europe, the US, and Asia Pacific. (Steinbock, 2003)

In recognition of the emergence of a more collaborative model of socio-technological behaviour in emerging high-technology sectors, the OECD in its 2011 GII report entitled "Stronger Innovation Linkages for Global Growth", emphasises the need to develop a framework from which to measure the number and type of collaborative innovation linkages that are occurring at a system level, asserting that at the present time,

> The challenge is to detect and quantify the dynamic and often informal nature of linkages and their efficacy. A key measurement problem is that a significant share of collaborative activity remains unmeasured ... Innovation indicators for less-developed economies ought to assess the extent to which connections and linkages are present in the field of innovation, define the nature of these links – including determining whether they are national or international – identify involved or excluded agents, and ascertain the efficiency of existing information channels. (OECD, 2011)

Indeed, it has been asserted by the OECD, that the category of "technological cooperation" is the most useful for developing the next generation of globalisation indicators:

> There are three reasons to believe a focus on cooperative arrangements – innovation networks – holds this promise: First, innovation networks are becoming the major scientific and technological actors in the process of globalisation. Second, innovation networks provide critical information about the other two categories of globalisation indicators – technological generation and exploitation. Third, innovation networks provide a useful way to integrate input and output indicators of technological advancement. (OECD, 2011)

It is important to emphasise here that, due to the fact that intangible assets and relational capacity are the key competitive currency in the collaborative economy, ecosystem participants need to engender significant trust in order to be considered for ecosystem membership and alliance partnerships. Hence, it can be argued that in order for the Chinese state and its key domestic firms to achieve ecosystem membership and shaping capacities in the global socio-technological system, China's weak intellectual property regime will need to be urgently addressed.

Open innovation, patent tradability and China's weak intellectual property regime

It is widely held that the existence of strong intellectual property rights (IPRs) encourages innovation, which subsequently results in increasing benefits for everyone. The central thrust of the argument is that if strong property rights provide incentive structures for the production of goods, there must also exist appropriate incentives for the production and dissemination of ideas (Goldin and Reinert, 2007:204). However, Chinese firms have exhibited a persistent lack of respect for IPRs. Indeed, China is the world's single largest producer of pirated goods. China possesses an extensive and sophisticated network of intellectual property infringers who pirate products from virtually every industry imaginable. For example, China is one of the world's largest manufacturers of fake DVDs, designer goods, car parts and pharmaceuticals. It has been estimated that 90–99% of software in China is pirated. Furthermore, counterfeit labelling of agricultural and other food products is common. Moreover, counterfeiters have also even begun to copy the experience of shopping itself by replicating the storefronts and interiors ("trade dress") of famous global stores and restaurants, including IKEA, McDonald's, Starbucks, Dairy Queen and the Apple Store (Kassner, 2012).

This has led to low confidence in the Chinese IPR regime, uncertainty in the rule of law, and bureaucratic and complicated human resource management. Indeed, the lack of IPRs is often cited by business representatives as a barrier to R&D in China. For example, in a 2004 survey conducted by the Economist Intelligence Unit (EIU), 84% of executives cited inadequate IP protection in emerging economies as a challenge for R&D globalisation. According to Hagedoorn, the protection of IP is a fundamental component in international partner selection and related R&D behaviour (Zhao, 2010).

In a similar vein, firms in the IP-intensive US economy are increasingly being deterred from entering R&D or co-development alliances with Chinese firms due to the significant losses that this has incurred for firms. In 2009, losses of approximately $48.2 billion in sales, royalties or license fees due to IPR infringement in China were reported. This estimate falls within a broad $14.2 billion to $90.5 billion range because many firms were unable to calculate exact losses. Of the $48.2 billion in total reported losses in 2009, approximately $36.6 billion (75.9%) was attributable to lost sales, while the remaining $11.6 billion was attributable to a combination of lost royalty and license payments as well as other unspecified losses (Commission on the Theft of American Intellectual Property, 2013). In 2010, Commander of the U.S. Cyber Command and Director of the National Security Agency, General Keith Alexander, stated: "Our intellectual property here is about $5 trillion. Of that, approximately $300 billion [6%] is stolen over the networks per year". He later called the theft "the greatest transfer of wealth in history" (Commission on the Theft of American Intellectual Property, 2013).

What needs to be emphasised is that because contemporary high-technology firms have few physical or tangible assets, as already established, a significant component of their value is derived from their intellectual property portfolios. As a direct result, in high-technology sectors, where knowledge is a firm's primary asset for generating competitive advantage, the formal protection of pre-existing knowledge and intangible assets via formal legal mechanisms such as patents and trademarks is an important strategy for firms wishing to engage in collaborative alliances for future technological knowledge generation. Indeed, it has been argued that it is no accident that we are beginning to see a simultaneous increase in open innovation, markets for technology, and the importance of IP in firm strategy, primarily in the form of patents and cross licensing (Hall, 2010).

At first glance the two concepts (open innovation and IPR protection) appear to be irreconcilable. Open innovation implies a willingness to allow knowledge produced within the firm to spill over to others (possibly with the expectation of receiving knowledge spillovers from others in return). In contrast, IPR protection enables a firm to exclude others from using that knowledge. It has been argued that the latter creates a "patent fence". A patent fence is used to refer to the idea that stronger patent rights can have an adverse effect on innovation and stifle – not facilitate – innovation and knowledge transfers. However, unlike the closed innovation approach which regards patents as monopolistic "fences" and "barriers" to keep competitors away, an open innovation approach considers patents as "currency" that can be used to acquire access to third-party IP to expedite technology development through non-exclusive licensing and/or cross-licensing (Hall, 2010). Furthermore, the open innovation approach allows for the selling or bartering of non-essential IP to secure at least some return on the investment for IP protection, instead of simply abandoning patents that have little or no direct value to the IP holder (Hall, 2010).

Conversely, it can be argued that in contemporary technological innovation ecosystems patents are conceptualised not as exclusionary mechanisms per se but as a tradable currency that provides firms with a valuable bargaining tool from which to co-develop technological assets and products and gain access to technological ecosystems. This means that the Chinese government and its domestic firms need to build up a stock of core technological patents as well as develop an effective IP protection system for both domestic and foreign firms.

Policy directives and recent governmental press releases indicate that the Chinese government is beginning to acknowledge the growing importance of IP reform as an important prerequisite for high-technology ecosystem development: "Only through effective intellectual property protection can science and technology innovators be rewarded for the fruits of their labors ... By doing so, their rights and innovative spirit can be protected and the momentum can be maintained. Protecting IPR is a necessary means to maintain market integrity and encourage enterprises to invest more" (Hu, 2010). In November 2011, after a nine-month special campaign against IPR infringement and counterfeits, the State Council, headed by Vice Premier Wang Qishan, set up a leading group against IPR infringement and counterfeits. Two specific campaigns were launched to address the concerns of foreign firms, including "Campaign Thunderstorm" to fight against patent infringement and counterfeiting, and "Campaign Skynet" to fight against patent fraud. China now seems convinced about the importance of patents: "As a fundamental system to encourage and protect innovation, the patent system is playing an increasingly important role in economic, technological and social development of a country", says the State Intellectual Property Office document that calls the patent drive an "inexorable requirement to deal with fierce international competition" (Murthy, 2011).

An example of the Chinese government's commitment to reform its IPR system can be drawn from a number of recent cases where foreign firms were awarded damages for IPR infringement by a Chinese firm. These include a case by a German company that won a settlement of $3 million in Beijing for infringement of its design for a bus, a British company that sued successfully over the heating element of a kettle and a firm from Wuhan that won $7 million in a case against a company from Fujian and its Japanese supplier over the use of a process to clean sulphur (The Economist, 2010).

What needs to be emphasised here is that the more Chinese businesses turn to innovation and services rather than manufacturing, the more they will become active producers and stakeholders in IP reform. Indeed, Chinese businesses have already started to demand reforms to strengthen IP enforcement within China, and a handful of businesses have utilised foreign courts to protect their intellectual property rights abroad. As China continues to transition from a developing nation to a more sophisticated global economy, the stage is being set for incentives to favour stronger IP enforcement (Evans, 2003). This IP regime will need to take into account newly emergent forms of collaborative IP protections, discussed earlier in this chapter, as well as more traditional IP protection mechanisms.

Alliance capitalism, resource sharing and relational ties

As it has been established in this chapter, the significant increase in cross-border collaborations over the last two decades is one of the most prominent components of the technological globalisation process. Indeed, collaborative initiatives have become the preferred institutional form through which diverse forms of development initiatives are designed and implemented. "That is global development partnerships and collaboratively-governed standards initiatives are the institutional pathway of choice for a new generation of initiatives dealing with everything from global health to the marginalisation of so-called 'blood diamonds' to the role of telecommunications in empowering peasant farmers in global markets" (Zadek, 2006).

For example, evidence of this shift in the US can be seen by looking at *R&D Magazine*'s R&D 100 Awards. For 45 years, R&D Magazine has been recognising the 100 most innovative commercial products introduced in the previous year. In 1975, 47 out of 86 domestic innovations were produced by Fortune 500 companies, and 40 of these involved no outside partners. By 2006, the Fortune 500 companies were responsible for only 6 out of 88 innovations, and in most cases, partners were involved (Block, 2008).

The extant literature on alliance capitalism contains relatively little research on the development of alliance capitalism in China or on the role of the state in the facilitation and maintenance of such alliances. Alliance capitalism is a term used to describe the increasing interdependence of economic entities and the partial erosion of hierarchical control over value-added activities in favour of network-based collegial entrepreneurship (Dunning, 1997a). Primarily, the existing literature is pitched at the level of the firm with an overt focus on the nature of its internationalisation. Hence, this literature fails to address the dynamic and highly interdependent nature of high-technology ecosystems and knowledge zones or the role of the state as a network organiser and system shaper. In order to address this theoretical gap we need to reframe current conceptualisations of alliance capitalism, with its overt focus on the internationalisation of the firm, to incorporate the notion of reciprocal dependence whereby firms and states from differing global locations are encouraged to form alliances; not just to function effectively in the global economy and gain market access, but also to address the existence of critical ecosystem dependencies and the need for interoperability and convergence. As a consequence, the role of the state in the development process is transformed as it endeavours to revise its governance structures and policy directives in a way that is responsive to global-national manifestations of economic and technological embeddedness.

It is important to emphasise here that the concept of network legitimacy, reciprocal independence, relational networks and collaborative governance are all important contemporary variables that operate to mould and define the parameters of the contemporary alliance-building and bargaining environment in the global wireless communication sector and hence provide key analytical variables for theoretical analysis. Network legitimacy and relational networks refer to the process of cooperation and competition between network constituents. Over time, and over many exchanges and interactions, these orientations are linked to sanctioned relational behaviour – all actions in the industrial network system. That is, they form sets of relational ties and rules of behaviour and exchange (Dicken et al., 2001). In China's wireless communication sector, network justification and the behaviour of firms in the network emanates primarily from the development and commercialisation of innovative and creative technological solutions. This requires creative management of technical resource and activity links, connecting a firm's technology to other firms in both the domestic and global network. As a result, cooperative and competitive market, equity, research, software, and training alliances and ventures are formed between network actors. For example, in order to gain network legitimacy in a Chinese context, telecommunication giants like Motorola, Nokia and Lucent have engaged in resource- and activity-sharing programs with local Chinese equipment manufacturers. Active and constructive participation in programs like joint manufacturing, marketing, technical alliances, and contributions to social causes and regional developments have earned these companies network legitimacy and facilitated relational interdependence among significant local network constituents, including local customers, suppliers, equipment manufacturers, carriers, and provincial and central government authority (Low, 2007 cited in Low, Johnston and Wang, 2007).

Collaborative governance, as it is termed, is a mechanism designed to bring public and private stakeholders together in collective forums with public agencies to engage in consensus-oriented decision making. Defined by cooperation and win-win incentive structures for stakeholders the overarching goal of collaborative governance is to expand participation, redefine power and resources imbalances, and construct mutually beneficial leadership and institutional designs. Key factors that determine the success of the collaborative process itself include face-to-face dialogue, trust building, and the development of commitment and shared understanding.

Relational assets are the mechanisms with which an actor forms and governs relationships with other actors. Firms and individuals use their relational assets to gain access to other actors' assets and to coordinate the use of their partners' assets with the focal actors' own resources. The basic idea here is that the ability to leverage resources that other actors control arises from the ability to engender trust in one's own judgement and intentions. Relational assets provide ownership advantages through superiority in coordinating the use of functional assets (Mitchell, 2003).

It is via this interaction with foreign firms that domestic Chinese manufacturers have gained access to global markets, equity and technology processes. Thus, it can be seen that in an industrial network system, network legitimacy occurs in a context of reciprocal interdependence of both a complementary and substitute nature. It involves both collaborative governance and alliance development that in a Chinese context is, to a certain point, state directed. By "state directed" I mean that the Chinese government plays a fundamental role in providing funding and long-term anticipatory policy planning in a strategic attempt to respond to the emergence of technological ecosystem dependencies and a more collaborative global innovation system. Its fundamental goal here is to maximise the participation and positioning of its domestic firms in globalised high-technology ecosystems.

The key point to note here is that in the contemporary socio-technological system, technological ecosystems where critical ecosystem dependencies exist, actors are no longer focused on the ability to govern and control other actors' assets in order to enhance and prolong the value of their own intellectual assets. Rather, they are focused on forming alliances and co-development models that ensure the successful appropriation and creation of value by all ecosystem actors. The type and structure of alliances that occur in the emerging collaborative economy differ in a number of important ways from those that occur in a more closed economy or economic sector. In earlier closed economic models, strategic alliances took the form of joint ventures, R&D partnerships, technological alliances, licensing, franchising and cross-manufacturing arrangements. Among these, the most popular form of a negotiated alliance is the joint venture where two firms provide equity shares towards a company with a specific goal (Dunning, 1997). In the collaborative economy, technological and organisational alliances are designed to accelerate the innovation process itself, address the existence of critical ecosystem dependencies, and develop the technological and learning capacities of its members. This involves the development of new technological ecosystems, alliance facilitation, IP sharing, and the construction and coordination of new markets and forms of economic cooperation (Han et al., 2012).

These new alliance forms can be termed innovation ecosystem-building alliances. Such alliances differ from traditional, closed alliances in many aspects, including their strategic scope and scale, governing mechanisms and member composition. It is important for the state to understand and assess the potential value inherent in these new modes of collaboration. Firstly, unlike traditional alliances, innovation ecosystem-building alliances transform dynamically over time as new members join. Although studies abound regarding the market reaction to strategic alliances at the time of formation, little is known about the ongoing value creation and wealth spillover that result from changes in member composition over time. Secondly, the main goals and value drivers of innovation ecosystem-building alliances are to increase overall market demand rather than market share. In other words, ecosystem-building alliances seek to enlarge the "economic pie" through value co-creation, rather than fighting with competitors over a "fixed pie". Consequently, the rivals who do not participate in the alliance may also benefit from the innovation-ecosystem-building alliances (Han et al., 2012; Lee, Olsen and Trimi., 2012; Piller, Ihl and Vossen, 2010; Mitleton-Kelly, 2011; Roser, DeFillippi and Samson., 2013).

For example, the alliance goals in global innovation networks are often focused on the creation of new, revolutionary products (e.g., smartphone devices) or services (e.g., interactive digital TV services). These new products and services are enabled by technological innovations that are co-developed by innovation ecosystem-building alliance partners and can enhance the industry's total profitability, amplifying the business prospects of both participants and their rivals. For instance, this innovation-driven, open-collaboration model is frequently used by firms operating in the mobile phone industry, including handset manufacturers, software developers and mobile operators. These collaborations facilitate value co-creation through the joint design and development of technologically innovative devices, services and standards. The collaborative manoeuvres currently being harnessed in high-tech industries possess the potential to reshape the competitive dynamics and alter the strategic positioning of the companies that operate within this vibrant and fast-paced environment. Moreover, the technological innovations cultivated through innovation ecosystem-building alliances could enable the participating firms to develop and introduce an entirely new market, which would create substantial economic value and opportunities for all parties involved in such collaboration (Han et al., 2012).

Furthermore, in an open innovation model, a company commercialises both its own ideas as well as innovations from other sources, leveraging its traditional market pathways but also seeking new ways to bring these ideas to market via new, external company pathways. Under the open innovation model, there are many ways for ideas to flow into the process and many ways for it to flow out into the market. Projects can be launched from either internal or external technology sources, and new technology can enter into the process at various stages. Note that the boundary between the company and its surrounding environment is now porous (like a colander), enabling innovations to move more easily between the two. In many industries the logic that supports an internally oriented, centralised approach to R&D has become obsolete. Useful knowledge has simply become too widespread, and ideas must be shared and harvested (Chesbrough, 2012).

From this theoretical perspective, globalisation profoundly alters the institutional environment of business and states, rendering the traditional two-tier firm–government bargaining model obsolete (Gardberg and Fombrun, 2007 cited in Windsor, 2007). This raises the questions in relation to the role of the state and multinational enterprises (MNEs) as key units of analysis (Schepers, 2007 cited in Windsor, 2007). This is because high-technology development is itself moulded and shaped by multiple actors in multiple countries at multiple levels in multiple policy arenas (Brewer, 2007 cited in Windsor, 2007). For example, the development of wireless communication standards involves national, regional and global bodies including the International Telecommunication Union

(ITU), 4G Americas, European Telecommunications Standards Institute (ETSI), Global System for Mobile Communications (GSM) Association, Global TD-LTE Initiative (GTI), Net!Works, Open Mobile Alliance (OMA), Small Cell Forum, TD Industry Alliance (TDIA), TeleManagement Forum (TM Forum) UMTS Forum (UMTS), Wireless Industry Partnership (WIP) and the Wireless World Research Forum (WWRF).

Conversely, it can be seen that new forms of state-led global alliance capitalism take the form of complex combinations of multiple actors and firms linked to one another through dense overlapping networks of economic and political connections. Any economic activity occurring in these networks can be said to be embedded simultaneously in multiple states, through both territorial and non-territorial channels. Certainly, alliance capitalism as reciprocal interdependence, it can be surmised, both signifies and demands a fundamental shift in the organisational behaviour of both multinational firms and governments. Theoretically, it poses an array of important questions in relation to the construction of the global economy and the institutional and relational aspects of state capacity.

Conclusion

It is becoming increasingly clear that the nature and pace of contemporary technological change and the structural changes to the contemporary socio-technological system at both a domestic and global level requires governmental policymakers to engage in a process of continuous adaption and self-organisation in order to respond to the dynamic, fast-paced and networked nature of the emergent socio-technological system. It is also clear that in evolving highly distributed innovation systems key roles exist for the coordination and linking of actors and knowledge assets.

Furthermore, the fact that innovation ecologies and spatially unconstrained innovation systems transcend national boundaries and are hence influenced by the policy jurisdictions of multiple states and technological knowledge zones and spaces indicates that the governance of complex socio-technological systems urgently requires the development of new sets of policy directives and capacities that address the globalised and spatially diverse nature of contemporary technological development processes. Specifically, the occurrence of networked innovation and collaborative alliances require strategic policy responses in the areas of resource and knowledge management and coordination, domestic- and international-level bargaining, alliance construction, and collaborative governance mechanisms.

What is important to emphasise at this point is that in the global knowledge economy bargaining power, resource and cluster development, coordination mechanisms, and the development of regulatory capacities are key policy attributes that serve to frame the construction of successful development plans. This is because the key role of the state in this environment is to incorporate new social groups, associates and organisations into bargaining processes and to find the right balance between various particularistic interests. This requires the development of collaborative architectural planning and coordination mechanisms.

In the following chapter, I will develop a framework for conceptualising the Chinese developmental state as a strategic actor committed to high-technology upscaling. I will explore its ability to respond to contemporary socio-technological system change at both a domestic and global level as well as its ability to adapt and engage with the dynamic, fast-paced and networked nature of the emergent socio-technological system. This will involve an examination of exactly what policy tools and alliance capabilities the Chinese government is developing and its capacity to influence and shape the contemporary socio-technological environment in a way that is conductive to its high-technology developmental goal.

In order to assess whether or not socio-technological systems change is shaping the policy directive of the Chinese government in the global wireless communication sector an empirical evaluation methodology needs to be established. As it has already been established in this chapter, socio-technological systems change in the global wireless communication sector is increasingly defined by sets of critical ecosystem dependencies and collaborative alliance structures for product development. Hence, Chinese firms and government participation in "collaborative alliances for product development" is a key analytical variable that can be employed to measure the extent to which socio-technological systems change is shaping the behaviour of both Chinese domestic firms and governmental policymakers. Another important variable that can be used as an analytical benchmark is the use of "trans-territorial governance" and "market-making mechanisms". That is, are Chinese firms and governmental policymakers in the sector engaged in trans-territorial governance, new modes of value appropriation and market-making behaviour? If it is found that the Chinese government is utilising these new developmental mechanisms, it provides empirical evidence for the conceptualisation that socio-technological systems change is playing an important role in both Chinese domestic firm behaviour and the development and execution of Chinese policy making in the sector.

4
Technological Development, Alliance Capitalism and Chinese State Capacity

Innovation is a catchword that serves to mould and define the strategic policy objectives of the twenty-first century Chinese state. The central government's guiding policy documents for the promotion of innovation is the Medium to Long Term Plan (MLP) for the Development of Science and Technology (S&T) which was issued in January 2006 and the 12th Five-Year Plan for 2011–15. These S&T documents articulate the key goals and policy directives that the Chinese developmental state will employ in its attempt to become an "innovation-oriented" society by the year 2020 and a world leader in S&T by 2050 (Bute, 2013). In 2012, the State Council issued another important policy document designed to accelerate the innovation process: "The Development Plan of National Strategic Emerging Industries during the 12th Five-Year-Plan Period (2011–2015)" (Yi, 2012). The goal of this policy document is to specify strategies and plans of nurturing and developing strategic emerging industries during the 12th Five-Year Plan. According to the State Council, strategic emerging industries are defined as sectors that are based on major technological breakthroughs, address major development challenges, have long-term impacts on economic and social development, are knowledge intensive and have growth potential (Yi, 2012). It is a state-led policy approach that establishes a central role for the government in the development and the steering of the innovation process. It is also defined by the way in which Chinese state-owned enterprises are becoming active global investors and participants in international and technology markets with increasing global market share in key sectors.

The ability of the Chinese government to upgrade the Chinese nation's competitive position in the global economy by utilising a highly interventionist state-capitalist model of economic and technological development is a point of much theoretical contention. Specifically, the idea that the state should play a leading role in steering and financing innovation policy has been the subject of fervent debate. In *China 2030*, a recent analysis of the Chinese economy, the World Bank asserted that the country requires: "a better innovation policy, [which] will begin with a redefinition of government's role in the national innovation system... [and] a competitive market system" (World Bank, 2013). US business executives go a step further and compare China's new innovation development plan to "the Borg in *Star Trek* – an enormous organic machine assimilating everything in its path, in this case the inventions of other nations". China will build its dominance by "enhancing original innovation through co-innovation and re-innovation based on the assimilation of imported technologies" (McGregor, 2010).

The key objective of this chapter is to present a theoretical and empirical overview of state capitalism as a developmental model for globalised innovative upscaling. Exactly how the Chinese state understands and responds to socio-technological innovation, its ability to develop indigenous knowledge assets, to insert itself and its domestic firms into emerging technological ecosystems and to facilitate global technological alliances in the emerging collaborative economy is of key interest to both the nation states that make up the international economic sphere and the international institutions designed to regulate it. This is because an analysis of the Chinese state's capacity to influence and shape the global socio-technological environment it is immersed in provides critical information regarding the future development of China's innovation system, its global reach, the effectiveness or rigidity of the institutions that define the international regulatory environment, the ways in which markets and techno-economical systems are embedded and dependent upon political and social systems, a synopsis of the most pertinent problems faced by late developing nations and an agenda for future research.

The developmental state

The relationship between the state and market under conditions of globalisation is one of the most controversial topics in the political economy. Specifically, the question of whether or not the process of technological and economic globalisation impinges on the capacity of developing nations to design and implement development strategies is the focus of much theoretical discourse from theorists at each end of the ideological spectrum.

In contemporary literature a latecomer nation state that possesses the ability to transform its economic policy prescriptions and policy tools to respond to global economic challenges and opportunities is termed a developmental state. Developed as a response to the failure of both neoclassical economics and dependency theory in explaining the rapid economic growth experienced in East Asia, the developmental state model has become a leading paradigm for understanding East Asian economic growth (Johnson, 1982).

Essentially, the developmental state paradigm is comprised of a collection of theoretical assertions and empirical descriptions that relate economic growth to institutional structures and organisational forms (Boyd and Ngo, 2006:1). In the literature the "developmental state" comprises two components: one ideological and one structural. It is this ideology-structure nexus that serves to distinguish the developmental state from other states (Ukwandu, 2009). In terms of ideology, a state is defined as "developmentalist" when it conceptualises its priorities and goals as that of ensuring economic development primarily interpreted as meaning high rates of accumulation and industrialisation (Ukwandu, 2009), and in a contemporary context, high-technology innovative development. That is, such a state's legitimacy rests upon its ability to promote high, sustainable economic growth and structural change in the productive system in both the domestic and international economy (Castells, 1992:55). It is a state that intervenes intensively in the economy by either direct involvement in the economy through state-owned enterprises, or by using policy tools to prioritise strategic industries with cutting-edge technology by "picking winners" or creating institutional settings designed to put the country in an advantageous position (Deyo, 1987; Gereffi and Wyman, 1990; Gilpin, 2001; Wade, 1990).

Gerschenkron's (1962) theory of "late industrialisation" is the earliest prototheory of the developmental state (Dent, 2004:80). In his seminal study on economic backwardness entitled Economic Backwardness in Historical Perspective, Gerschenkron emphasised that the fundamental developmental role of the state is the mobilisation and coordination of resources necessary for closing the techno-industrial gap between "late industrialising" and the advanced industrial nation states (Dent, 2004:80). Operating from documentary evidence drawn from the experiences of nineteenth-century continental Europe and the early-twentieth-century Soviet Union, Gerschenkron illustrated the importance of the state's industrial and mercantile policies in gradually strengthening a weak business sector in the pursuit of techno-industrial catch-up (Dent, 2004:80).

Gerschenkron's theory was further developed by Chalmers Johnson in his study of the Japanese Ministry of International Trade and Industry (MITI). Basing his observations in the model economic planning bureaucracy in Japan, MITI, Johnson devised a Weberian ideal type of an interventionist state that was neither socialist nor a free market. He termed this state variant as a plan-rational capitalist developmental state; a state that conjoins private ownership with state guidance (Johnson, 1982).

If the view of development as a process of sequentially mastering comparative advantage in technologically more sophisticated industries is correct, developmental states must also assume an active role in industrial policy and in stimulating the acquisition of dynamic comparative advantage. From this frame of reference, markets alone are insufficient to climb the ladder of comparative advantage, since the successful generation of new comparative advantage requires coordinated and anticipatory changes in investments, institution-creation and reform; and trade and tax-cum-subsidies policies. Only governments, it is argued, are capable of correcting coordination failures in a complex and ever-changing economic and technological system. In developed countries, where financial institutions and entrepreneurs are ready to engage in risky activities, markets alone may be adequate for the task...but not in developing countries where financial institutions and entrepreneurship are inadequate and institutional change can only be promoted by government (Anderson and Gunnarsson, 2003). However, as the 2008 US financial crisis highlighted, the state also has a role to play in overseeing the use of financial instruments that are risky in order to prevent financial loss and meltdowns caused by excessive speculation.

The Chinese developmental state

The Chinese state possesses three key institutional features identified by developmental state theorists as central characteristics of a developmental state: transformative goals, a relatively insulated bureaucracy and institutionalised government-business relationships (Deans, 2004:134). However, whilst a large academic literature devoted to examining the locus, implications and international ramifications of China's economic transformation exists, it is largely devoid of analysis framed by the wider research agenda pertaining to the developmental state (Deans, 2004:134). This is primarily because the Chinese developmental state differs in a number of fundamental ways from other developmental states.

In order to address the institutional and theoretical anomalies posed by the Chinese state, Ming Xia in his book, The Dual Developmental State (2000), constructs a modified developmental state paradigm that asserts that the Chinese developmental state is dual in nature, sustained by both the legislative and the political institutions of the Chinese state (Xia, 2000:26). One of the most distinctive aspects of the Chinese reform process is its adoption of a dual-track system. The dual-track system refers to the coexistence of a traditional plan and market channel for the allocation of a given good (Naughton, 2005:8). Instead of quickly dismantling the planned economy, the reformers decided to facilitate the functioning of the planned economy in order to secure stability and protect the status of key government initiatives in sectors such as energy and infrastructure. A two-tier price structure was implemented. For example, a single commodity will have both a typically low state-set price and a higher market-set price (Naughton, 1995:8). However, a fundamental feature of the Chinese economic transition is its commitment to the transformation of its economy from a command-controlled to a market-controlled one with the plan being replaced as the economy gradually grew out of it (Naughton, 1995:9).

From this theoretical perspective, there exists in China a unique dual structure that radiates from the central to the local levels, between the governments of the People's Congress (PC) (Xia, 2000:27). This system of the PC legitimates the ruling communist party, integrates the nation and ratifies the large number of laws required for market creation and the supervision of the judicial branch (Xia, 2000:27). Moreover, the dual development state is comprised of a two-tiered structure in which the developmental role of the central state is nestled within the context of a central/local state synergism (Xia, 2000:27). This allows reform-minded local leaders to experiment with new ideas and policies. However, a mandate for reform is required from the centre, which means an inter-related relationship between the centre and local state has been formed. As a direct result, the reform process in China is guided by interaction between central and local governments and a power-sharing mechanism that facilitates the mobilisation of the population and their commitment to the development agenda (Xia, 2000:27).

Methodologically, Xia utilises the theoretical and empirical concepts of comparative transaction cost analysis. Transaction cost analysis is a useful theoretical mechanism that can expose the transaction costs associated with the exchange of ownership rights to economic assets and the enforcement of exclusive rights. Its fundamental premise is that transaction costs include the cost of arranging a contract ex ante and monitoring and enforcing it ex post (Xia, 2000:27). That is, it is "the costs of running the economic system" (Xia, 2000:27). Comparative transaction cost analysis is usually applied at the level of the firm. From this frame of reference, the government is conceptualised as a super-firm that is involved in a process of economic and political cost benefit analysis in order to ascertain the specific set of costs associated with the implementation of a developmental agenda (Xia, 2000:27).

Furthermore, Xia extends upon the classical developmental state paradigm and incorporates Oliver Williamson's (1975, 1996) concept of "the mode of governance" into his theoretical model. As a theoretical concept, the mode of governance refers to economic and social transitions that are organised into three types of institutional matrices, or modes of government (Xia, 2000:4). These are market, hierarchy and hybrid (Xia, 2000:4). In the literature, most theorists agree that the Chinese developmental state model is neither market nor hierarchy but hybrid (Xia, 2000:4). That is, since the inception of the reform process in 1978, the Chinese state has transformed its political and economic structure to one that can no longer be defined as socialist nor can it be classified as capitalist. Phillip Deans calls this hybrid economic model a post-socialist developmental state (Deans, 2004:134).

Firstly, China as a post-socialist developmental state is engaged in a complex process of transition from a command to a market economy (Deans, 2004:134). Secondly, China has a large population base and vast geographical size (Deans, 2004:134). This means that coordinating an integrated and comprehensive economic development program from an authoritarian centre is complex and hard to monitor effectively. Thirdly, in China, Deng Xiaoping initiated the reform process with legislative decentralisation (Deans, 2004:134). Conversely, the reform process has resulted in a significant redistribution of economic decision-making power, primarily from central to local-level authorities. The central authorities have granted local governments across China varying degrees of financial autonomy. In addition, some have exploited growing international contacts to find further sources of extra-budgetary investment finance, or pressured theoretically (quasi) independent banks and enterprises to support local development plans. As a result, China's developmental trajectory owes at least as much to the dysfunctional agglomeration of numerous local initiatives, as it does to the plans and strategies of the national-level decision-making elites. This has resulted in a political control structure that has been termed fragmented authoritarianism. Fragmented authoritarianism contrasts sharply with the centralised bureaucratic authoritarianism that characterises and defines a large proportion of developmental states (Moore, 2002).

The developmental state's theoretical focus on the organisational structure of the state and its conceptualisation of the state – not as a single monolithic actor, but instead as a strategically important facilitator that coordinates political and economic strategies and provides resources and incentives designed to encourage industrial transformation – is compatible with an important body of theoretical scholarship on post-Mao China: the fragmented authoritarianism literature in which the state is no longer conceptualised as the monolithic authoritarian actor it was in both the totalitarian models or Mao- or Deng-in-command literature (Moore, 2002:279). What is important to emphasise here is that this fragmented political and economic institutional infrastructure raises important questions regarding the nature of the policy responses available to the Chinese state in responding to adjustment challenges.

The fragmented political and economic authority that defines the Chinese policy-making system is commonly characterised as a bargaining system (Moore, 2002:282). It is precisely these deep structural divisions within the bureaucratic system that conspire to mould and define the range of policy options available, as different policy networks advance and negotiate a range of policy choices. However, it has been argued that the fragmented nature of Chinese economic and political authority across the industrial landscape combined with a weakening of the formal planning system has resulted in a reduction of the ability of the state to respond effectively and authoritatively to administrative and economic development issues. That is, the Chinese state lacks the embedded autonomy which has been cited as a key variable by the developmental state literature required for successful development by a latecomer nation (Evans, 1995:32). "Embeddedness" means that economic actions are embedded in a network of concrete social relations (Granovetter, 1985). Whilst the original conceptualisation was employed as a medium to examine local and social interpersonal relations, recent studies have extended upon the concept and applied it to analysis of government institutions and policies, and the operations of the nation state itself (Sit and Lin, 2000:654).

Moreover, whilst current conceptual literature on the developmental state stresses the importance of planned industrial policy, the Chinese reform process, until 2006, is notable for its lack of a long-term plan and its commitment to experimental incremental change. That is, the reform process is defined by its lack of a blueprint and its emphasis on prior experimentation and learning. Thus, the fundamental results and overarching shape of the reform process are time and context dependent. The Chinese have termed this process "grouping for stones to cross the river" (Naughton, 1995:5). Yet, the fact that China did not possess a predefined economic plan and sequence of reform initiatives, it can be argued, has provided it with an adaptive edge that has allowed it to continuously monitor economic results and quickly adjust its policy prescriptions in an increasingly unpredictable, global, networked, economic environment.

It is important to note here that the co-production of innovation ecologies and the technological governance processes that transcend traditional territorial boundaries have important implications for the way the Chinese developmental state operates and the types of capacities that it needs to develop. The spatial extension of state capacity from one that is focused on national economic and technological development goals to one that is focused on the development of globalised technological ecosystems, and the fact that investment and growth in the contemporary knowledge economy is increasingly driven by intangible assets, means that the contemporary developmental state, committed to technological development and upscaling, is presented with a whole new set of developmental variables and objectives. Hence, traditional models of the developmental state need to be modified and extended upon in order to incorporate the key developmental objectives of states in an era of intangible assets and high-technology trans-territorial development.

From this theoretical perspective, the impact of globalisation on the capacity of the developmental state is defined not by the diminution of state capacity per se, but in the creation of new forms of policy capacity (Jayasuriya, 2004). Specifically, it is argued, traditional forms of state capacity are defined by the transformative model. This model is understood in terms of a particular set of attributes or "key endowments that a state or public agency may possess to give it a set of transformative powers over policy and structure" (Jayasuriya, 2004:488). These traditional forms of state capacity include infrastructural capacity, transformative capacity and distributive capacity. Whilst these types of state capacity are all important policy tools employed by the developmental state to achieve their developmental goals, the fact that contemporary technological governance is located in multiple sites, and involves a number of non-state actors, argues Jayasuriya (2004), means that it is necessary to reconceptualise state capacity as a form of socio-economic engagement rather than as a series of organisational attributes.

Two important new forms of state capacity, it has been argued, are relational capacity and regulatory capacity. Relational capacity is important for states that wish to play a key role in their nation's developmental process. These states need to be able to engage with multiple actors in multiple sectors and require relational legitimacy for ecosystem building and shaping purposes. From this theoretical perspective, economic, political and technological development requires the recruiting of actors and resources within a specific sectoral ecosystem or particular field of governance:

> much of the so-called post-Washington consensus articulated by proponents such as the former World Bank chief economist Stiglitz (2002) is based on the idea that economic reforms need to be made institutionally sustainable. For these theorists, the concept of relational capacity captures the essence of capacity building as part and parcel of a wider process of legitimacy building that stands in contrast to the technocratic top-down models of governance privileged by attribute models of capacity. (Jayasuriya, 2004)

Regulatory state capacity is necessary for governments that seek to constitute new arenas of governance and regulatory reach at multiple levels and in multiple sectoral environments (Jayasuriya, 2004). The ability to change the global regulatory environment and construct new forms of regulatory management bestows on such states technological ecosystem power in key technological sectors. Regulatory state capacity can be seen manifest in new forms of assertive legalism and constitutionalism in economic and political decision making. For example, central bank independence represents an important example of how new arenas of governance are being created by states through an increasing process of legalisation or proceduralisation (Jayasuriya, 2004).

Conversely, it can be argued that new forms of state capacity are emerging, and these new forms of state capacity are contributing to the emergence of a new form of the state. It is a state made of shared institutions and enacted by bargaining and interactive iteration all along the chain of decision making: national governments, co-national governments, supra-national bodies, international institutions, and governments of nationalities, regional governments, local governments and NGOs (Jayasuriya, 2004). Decision making and representation take place all along the chain, not necessarily in the hierarchical, pre-scripted order. This new state functions as a network in which all nodes interact and are equally necessary for the performance of the state's functions.

This new form of the developmental state has been termed the global network state (GNS).

The GNS, asserts Sean O'Riain, is defined by its capability to form networks for the production of technological innovation, and by its ability to attract foreign investment and stimulate domestic growth. It achieves these key goals by providing information about transnational corporations and world markets to domestic firms, and develops institutional frameworks and funding decisions designed to provide the necessary conditions for stimulating connections between domestic and international companies. That is, the developmental network state's overarching goal is to facilitate collaboration and encourage innovation at a both a domestic and global level (O'Riain, 2000).

In transforming itself to operate within a locally and globally networked economy and polity, asserts O'Riain, state governance itself is "re-scaled" as the prior privileged role of the national level gives way to a "glocal" form of state (O'Riain, 2000:165). Furthermore, coherent state bureaucracies may not be the only organisational structures which may promote embedded autonomy. From this theoretical perspective, embedded autonomy is not guaranteed by a coherent bureaucracy per se, but by the flexibility of the state structure itself. This is because the decentralisation of state agencies enables them to become deeply embedded in their local economies and their specific economic sectors, despite the fact that they are often dealing with a wide range of individuals and organisations across widely dispersed networks (O'Riain, 2000:165).

Moreover, the role of the state in achieving its development goals is not to assume the task of development itself but instead to shape the capabilities of the business sector, society and its associative markets to do it (O'Riain, 2000:166). This requires the building of local and global network capabilities and innovation systems. It is important that these systems and network structures are designed to be adaptive enough to operate at the forefront of the future regulatory, market and technological environment. In this way these emergent network structures and global alliances can acquire lucrative first-mover advantages (O'Riain, 2000:165).

However, as it has previously been established, in this new more collaborative environment, achieving first-mover status is not the only way to move up the technology ladder and acquire market share. Instead, ecosystem shaping, co-development and system embeddedness are new strategic models of behaviour that can be utilised by the developmental state to facilitate both the successful adoption of its national firm's products in the contemporary global market system as well as help to facilitate the co-creation and development of new technological products that are both domestic and foreign in ownership.

The contemporary Chinese developmental state exhibits numerous features of a GNS. Certainly, it is highly responsive to transnational technological innovation and has developed a range of policies designed to stimulate innovation-driven growth at both a domestic and a global level. Specifically, the Chinese state as a strategic actor has played a key role in facilitating both the development of indigenous technology as well as global alliances between its domestic and foreign firms. Furthermore, whilst the Chinese state is a highly effective bureaucratic actor with the power to exercise top-down policy-making decisions, it is also deeply embedded in a "network polity" that is comprised of an array of sociopolitical alliances that are local, national and global in nature.

The type of state capacities being utilised by specific nation states is an important analytical variable that can tell us a lot about their developmental path and capacity for high-technology upscaling in the contemporary collaborative economy. Hence, in the following sections of this chapter I will examine the policy tools and state capacity of the Chinese government as it attempts to become a global ecosystem shaper and regulatory actor in order to facilitate the technological development of the Chinese nation state.

Chinese state capacity and high-technology innovation

State capacity is itself a core concept in the political economy of development. It provides us with a powerful tool from which to isolate and study the process of technological and economic development. Primarily, it refers to the ability of the state to formulate and implement policy. Any analysis of Chinese state capacity must be framed by an understanding of the specific form of state capitalism that it practices. Essentially, state capitalism in China is defined both by the way in which Chines state leaders exert control over the development process and the way in which state-owned hybrid companies are responsible to both the state and the market. This form of state developmentalism has been termed state capitalism 2.0 (Musacchio, 2012).

It is important to note here that this model of state capitalism bears little resemblance to the planned socialist economies of the past. It is a form of state capitalism that is becoming increasingly used by developing nations to address the way in which contemporary high-technology assets are produced in global innovation networks, the emergence of intangible intellectual assets and the increasing existence of critical ecosystem interdependencies in high-technology sectors.

Moreover, state capitalism 2.0 can be conceptualised as a modified model of earlier developmental state forms that has sought to adjust to contemporary socio-technological systems change and develop a range of new state-interventionist policy tools that are designed to address the increasing need for the state to play a role in funding long-term risk capital to help facilitate the development of technologies with long time frames in emerging fields such as nanotechnology, green technology and biotechnology. It is also designed to provide a mechanism for the state to play a trans-territorial role in the coordination and shaping of technological ecosystems at a global level.

This new hybrid form of state capitalism, it is argued, is not defined by any single political or economic model per se, but by the way in which the state uses its power strategically to facilitate economic development. In this new model of state capitalism, hybrid companies are responsible to both the state and the market, and profitability and technological development are its key goal. This hybrid form of capitalism – state support disciplined by the market – gives state capitalism three huge advantages, according to Musacchio: "It produces global champions that have quickly risen up the ranks of the world's top companies. It gives companies the freedom to invest for the long-term rather than obsessing about short-term profits" (Musacchio, 2012). Indeed, it has been argued that increased external infrastructural spending that is financed via pools of state capital held by sovereign wealth funds and direct investment by state-owned enterprises and private firms are in the process of facilitating significant global structural changes (O'Brien, 2014).

Certainly a highly distinctive characteristic of state capitalism in China is the central role of approximately 100 large, state-owned enterprises (SOEs) which are controlled by organs of the Chinese national government in critical industries such as steel, telecom, and transportation. China now has the world's third-largest concentration of Global Fortune 500 companies (61), and the number of Chinese companies on the list has increased at an annual rate of 25% since 2005. It is these companies that are China's national champions. Moreover, more than two-thirds of the Chinese companies listed on the Global Fortune 500 are state-owned enterprises. Excluding banks and insurance companies, controlling stakes in China's largest and most important of firms is owned by a central holding company on behalf of the Chinese people, the State-owned Assets Supervision and Administration Commission

(SASAC), which has been described as "the world's largest controlling shareholder" (Lin and Milhaupt, 2011).

Musacchio proceeds to divide state capitalism 2.0 into two main types: Leviathan as a majority investor and Leviathan as a minority investor. He writes:

> Closer to the more familiar view of state capitalism as a process involving outright state management, the state can act as a majority shareholder and manager of SOEs – a mode we refer to as Leviathan as a majority investor...In the Leviathan as a majority investor model, the government usually exercises control of SOEs indirectly, by appointing managers and boards of directors. In some SOEs, however, ministers act directly as presidents. Moreover, SOEs can be fully owned by the government or they can be publicly traded, as long as the government is the majority shareholder. Governments also exercise their control as a majority investor using large companies as conglomerates controlling a series of firms or through what is known as state-owned holding companies (SOHCs). (Musacchio, 2012, original emphasis)

The state can also influence the economy in an indirect way, asserts Musacchio, by acting as a minority shareholder and lender to private firms. Musacchio terms this mode of state capitalism "Leviathan as a minority investor". This model of state capitalism is a hybrid form, which mixes features of full state control and the private operation of enterprises. The key channels used by Leviathan as a minority investor are, it is asserted are: "holding shares in partially privatised firms (PPFs); minority stakes under state-owned holding companies; loans and equity by state-owned and development banks; sovereign wealth funds (SWFs); and other state-controlled funds (e.g. pension funds, life insurance)" (Musacchio, 2012).

It has been argued that state capitalism as an economic model is beginning to either replace or subsist with the free market. In response to the increasing use of state capitalist economic models by developing nations, and in order to facilitate debate regarding their potential impact on the global economy, in January 2012, The Economist published a provocative article entitled "Emerging-Market Multinationals: The Rise of State Capitalism", in which it stated that:

> the era of free-market triumphalism has come to a juddering halt....Liberal capitalism in the U.K. and the U.S. isn't just convulsed

with internal crises caused by unregulated financiers. It now faced, "a potent alternative": state capitalism, which has on its side one of the world's biggest economies – China – and some of its most powerful companies – Russia's Gazprom OAO China Mobile ltd. (CHL), DP World Ltd., and Emirates Airline. (Mishra, 2012)

Indeed, asserts Mariana Mazzucato in The Entrepreneurial State (2011), a strong case can be made that the role of the government, in the most successful economies, has gone way beyond creating the right infrastructure and setting the rules. Instead, it has been a leading agent in achieving the type of innovative breakthroughs that allow companies, and economies, to grow, not just by creating the "conditions" that enable innovation. Instead, the state can proactively create strategic policies around a new high-growth area before the potential is even comprehended by the business community (from the Internet to nanotechnology), funding the most uncertain phase of the research that the private sector is too risk averse to engage with, seeking and commissioning further developments, and often even overseeing the commercialisation process. In this sense it has played an important entrepreneurial role. As Mazzucato asserts: "The only way to make growth 'fairer' is for policy makers to have a broader understanding of the role played by the state in the fundamental risk-taking needed for innovation" (Mazzucato, 2011).

Furthermore, it is interesting to note that professional venture capitalists in the United States have concentrated their activities and earned their returns in a very small number of industrial domains. For instance, in the three decades since 1980, information and communications technology (ICT) has accounted for 50–75% of all dollars invested by members of the National Venture Capital Association (NVCA), with its average share usually hovering around 60%. The ICT and biomedical sectors together have consistently accounted for 80% of all dollars invested by venture capitalists. What is important to emphasise here is that it is only in these key sectors that the state invested at sufficient scale in the translation from scientific discovery to technological innovation (Janeway, 2012). That is, the only reason that venture capitalists saw such high returns in these sectors is because the early research and development process had already been funded by the state.

The central problem here is that venture capital and private equity funds, which are the primary funding mechanisms that private firms in the contemporary knowledge economy utilise to obtain funds to conduct R&D and fund the commercialisation processes, are defined by a short termism that is highly antithetical to deep innovation. As

Richard Jannaway articulates: "They're the invisible hand of capitalism itself: reaching out for quick gains while avoiding as much downside risk as possible. Private markets are loath to put up the billions of dollars and patiently wait a decade or more, yet that's precisely what's needed to develop breakthrough drugs, disruptive clean energy technologies or that new, new thing you and I can't yet imagine" (Jannaway, 2012). Furthermore, Jannaway and McKenzie analysed venture capital funds from the 1980s until the post-bubble era. Their research found a strong correlation between the level of venture capital investment and the occurrence of initial public offerings (IPO). For example, the median rate of return for the sample under favourable IPO conditions was 76%. In stark contrast, the medium rate of return for the sample under unfavourable IPO conditions was only 9%. This data supports the hypothesis that venture capital funds are primarily focused on short-term returns (Jannaway, 2012).

Thus, it can be argued that governmentally funded R&D programs and commercialisation support is especially important in a contemporary context due to the occurrence of long time horizons in some high-technology sectors for basic research and the significant decline in venture capital that has occurred following the 2008–11 financial crisis. For example, China's venture capital (VC) and private equity (PE) firms have seen a sharp drop in activity recently, following a period of rapid expansion. As a direct result, an industry reshuffle is now expected to occur, with many firms expected to shut down or seek opportunities in other sectors. Empirical data indicates that in the first three quarters of 2012, domestic PE firms raised a total of $7.9 billion for investment, which is a significant drop from the $35.7 billion raised in the same period of the previous year (Global Times by consultancy firm Zero2ipo) (Qian, 2012). Hence, a number of small-sized PE and VC firms, which were set up only two to three years ago, have already closed or turned to other business areas, asserts Li Weidong, a research director at the financial consultancy firm China Venture (Weidong, 2012 cited in Qian, 2012).

Furthermore, it has been conjectured that the combination of authoritarian hierarchy and collaboration within the high-powered incentive structures that define the Chinese state's bureaucratic infrastructure is reminiscent of another capitalist mechanism of transitions: private equity investments. From this theoretical perspective, the Chinese pattern of decentralised experimentation and innovation bears close resemblance to key features of the venture capital model as practiced in the United States (Lin and Milhaupt, 2011). For example, a key policy program recently initiated by the Chinese state is investing approximately 9 billion yuan ($1.32 billion) of government funds into new venture capital funds designed to support the growth of the Chinese high-technology sector. The state government has stated it will not take more than a 20% stake in each fund and will not exercise any control over the funds or firms so they can develop along market lines. The venture capital funds will be required to invest in early-stage companies and those with the potential for fast growth. Asserts the National Development Reform Commission (NDR): "The purpose is to promote the country's structural economic adjustment... and meet the broader goal of encouraging the development of high-tech industries" (NDR, cited in Lin and Milhaupt, 2011).

Certainly, it can be conjectured that innovation policy and R&D funding is exactly where state capitalism can play a fundamental role. However, the type of state intervention that state capitalists are conducting in the contemporary globalised socio-technological environment differs in a number of fundamental ways from previous rounds of technological development. For example, the state-led development strategy being employed by the Chinese government is highly adaptive and involves both system-structuring and system-shaping properties. Whilst the Chinese state intends to play a key strategic role in supporting the high-technology development and innovation of its domestic firms in sectors where the risk of investment is high, it also intends to do this in a way that responds to the need for collaborative behaviour and system integration at a global level.

National innovation and Science and Technology policy objectives and reforms

The Chinese leadership conceptualises innovation as an essential tool to promote its economic growth, maintain political stability, support advanced military capabilities and retain its global trade and geopolitical power. Ma Kai, minister of China's National Development and Reform Commission in 2006, argued:

> China's economic growth largely relies on material inputs and its competitive edge is to a great extent based on cheap labor, cheap water and land resources, and expensive environmental pollution. Such a competitive edge will be weakened... with the rising price of raw materials and the enhancement of environmental protection. Therefore, we should enhance [our] independent innovation

capability and increase the contribution of science and technology advancement to [our] economic growth. (Kai, 2006)

In short, for China, innovation is a policy of nearly unrivalled importance (Wolff and Ballantine, 2006).

China's current innovation development program is defined by its depth and scope, and the way in which it involves long-term planning beyond the single ministry technology development programs of the past and the way in which its execution is broadly dispersed between half a dozen ministries at the level of the central government (Wolff and Ballantine, 2006). The "Medium- to Long-Term Plan for the Development of Science and Technology" (State Council, 2006), is the third of its kind since 1949. This plan sets out the key objectives and priorities for the country's development in S&T as well as the main instruments that the Chinese government intends to use to achieve them. The overarching goal is that China becomes an "innovation-oriented" society by the year 2020 and – in the longer term – a leading "innovation economy" in the world by 2050 (Cao, Suttmeier, and Simon, 2006). More concretely, the objectives to be reached by 2020 are: China's R&D intensity will be increased to 2.5% of GDP (2.0% by 2010), innovation will contribute to 60% of economic growth and China's reliance on foreign technology will be reduced to below 30%; and overall, China will be among the top five countries worldwide in terms of key innovation output indicators (Fensterheim, Huang, and Murray, 2009).

In contrast to previous S&T plans, the latest plan specifically emphasises the need to develop capabilities for "indigenous" or "home-grown innovation", with a view to create the conditions for achieving a leading position in a number of S&T-based industries. These include the following: central government-initiated activities to promote indigenous R&D (applied at all levels of government); government funding of large-scale R&D programs and projects; preferential tax and financing policies to encourage domestic R&D; preferential government procurement policies to support domestic R&D; the development of high-value domestic intellectual property; and the designation of special economic zones that support domestic R&D activities with preferential access to infrastructure, financing and other services. The plan also seeks to develop domestic technical standards to decrease dependence on foreign technologies and increase the prominence of standards that rely on domestically controlled intellectual property. Incentives offered by local governments, such as tax exemptions and favourable terms for land use and utilities, reportedly are important factors that influence where R&D or production plants are located in China (United States Trade Commission, 2007). During the "17th National People's Congress" (NPC) which took place in October 2007, China's President Hu Jintao pointed out that the core of China's national development strategy is: "To enhance China's capacity for independent innovation and make China an innovative country...The Ministry of Science and Technology of China emphasises innovation as 'the soul of a nation'" (Fensterheim et al., 2009).

In addition, the S&T policy directives seem to reflect official enthusiasm to expand and develop future cooperation between foreign R&D firms and Chinese partners, and to improve prospects for further technology spillovers, with its emphasis on international cooperation and collaboration. It states that international S&T cooperation is an important aspect of China's policy of openness to the outside world. China is ready to enter into cooperation with any country depending upon the needs of Chinese S&T and economic development, according to the principles of equality and mutual benefit, mutual enjoyment of benefits, protection of intellectual property rights and respect for standard international practice. Cooperation can be bilateral, multilateral, with private parties or with governments in foreign countries at any level or through any channel (Irwin Crookes, 2008).

Hence, it can be seen that the plan contains two key strands. Firstly, there exists a drive to promote indigenous enterprise-level development to foster domestic intellectual property that can be licensed internationally and can reduce the country's reliance on overseas technology inputs (State Council, 2006). Nevertheless, the fact that the plan also emphasises the need for international cooperation and collaboration indicates that the Chinese government is currently engaged in the development of a delicate policy program that endeavours to combine the need for domestic technological and competitive upgrading with the need for international collaboration.

Key policy directives and innovation indicators

An examination of key innovation indicators such as patent data, R&D expenditure, number of published scientific papers and number of engineers and scientists suggest that so far the Chinese government's state-led innovation program is beginning to show results. Indeed, China's innovation development policy has begun to show a return in a number of key areas.

The capacity of the Chinese state to implement its innovation development agenda has been enhanced in the last decade by its accumulation of $3.2 trillion worth of foreign exchange reserves. Indeed, China now enjoys the world's largest current account balance. In 2011, it ran a $276.5 billion trade surplus with the United States (Atkinson, 2012). In 2007, the Chinese government set up the Chinese Investment Corporation (CIC), a sovereign wealth fund to invest and manage part of China's foreign trade surplus. As of August 2013, the fund had $572.2 billion in assets under management (Atkinson, 2012). The fund was set up to address the need to seek greater returns, increase diversifications and hold less US currency reserves. Since its inception, the CIC has made substantial investments in various asset classes, including direct investments and institutional real estate (Atkinson, 2012).

Furthermore, it is important to emphasise here that the Chinese development state is unique in that it possesses unprecedented power in its bargaining with foreign firms. In a survey conducted in 2004, the Delegate Office of German Economy interviewed 243 companies regarding their reasons for investing in China. Among the companies consulted, 94% cited future market access as a key reason for their investment in China. An additional 42% and 46%, respectively, cited low production costs and the necessity of following major customers as their reason for investing in China (Atkinson, 2012). Furthermore, with the 2008 financial crisis, the Chinese economy has now become a crucial engine and driver of economic growth. For example, in a survey conducted by the Economist Intelligence Unit (2011), almost half (49%) of all survey respondents indicated that the impact of the global financial crisis had raised their companies expectations for China. For larger companies with global revenues of more than $5 billion the figure is 73% (Economist Intelligence Unit, 2011).

A cornerstone of the Chinese developmental state's innovation development program is its commitment to increase its level of R&D expenditure. In 2011, China's R&D expenditure represented 1.97% of the GDP. In 2012, China further increased R&D expenditure by 17.9% a year to 1.02 trillion yuan ($162.24 billion) or 1.98% of GDP. By 2020, the Chinese government intends to increase this expenditure to 2.5% of GDP (Atkinson, 2012). Companies and enterprises invested the most in R&D, with 657.93 billion yuan recorded in 2011, up 26.9% year on year. Government-affiliated research institutes and universities spent 130.67 billion yuan and 68.89 billion yuan, up 10.1% and 15.3%, respectively, from a year earlier (Atkinson, 2012).

It is interesting to note here that China has overtaken the European Union in R&D expenditure as a percentage of GDP. The 28 member states of the European Union invested only 1.97% of their joint economic output in R&D in 2012, a mere 1% increase from a year earlier (Boehler, 2014). Japan reordered the highest level of R&D expenditure in 2012 at 3.4%, followed by the US and Germany who each spent 2.8%, respectively (OECD, 2012).

In 2012, the number of patents increased (albeit from a low level) by about 40%; China's share of total patents registered with the World Intellectual Property Organisation (WIPO), however, remains small. The majority of the patents are "design" or "utility model" patents. Nevertheless, in 2011, China became the world leader in the number of invention patent applications, with 217,000 applications filed, outperforming the United States, Europe and Japan (Zhen, Suny and Wrighty, 2012). In 2012, China secured its lead with the most patents granted. Moreover, for the first time ever, China's State Intellectual Property Office (SIPO) granted over a million patents. This increase in patent activity is a result of the Chinese central government's policy prescriptions that offer generous incentives for patent filing. For example, Chinese companies who file above a certain number of patents receive significant tax breaks. Tenure is more likely for university professors who are able to obtain patents. Patent application fees for qualifying individuals and companies are entirely subsidised by local government (Zhen et al., 2012).

The Chinese government has recently further extended its original IP patent policy prescriptions via its "Outline of the National Intellectual Property Strategy", released in 2008 and "Action Plan for Implementing National IP Strategy" (2009). The Chinese "National Patent Development Strategy (2011–2020)" "has been described as the first-of-its-kind intellectual property mission from any government" (Murthy, 2011). The plan asserts that by 2020, China will attempt to quadruple both patent applications in foreign countries and domestic patent applications for every one million people. The annual patents transaction financial target is to reach 100 billion yuan ($15 billion) by 2015. To boost domestic patents, China is offering incentives such as cash rewards, houses and tax breaks to scientists and innovators. Moreover, the patent policy directives promise infrastructure for quicker filing, examining and granting of patents, funding patent holders, and integrating new patents into the economy. Its patent proliferation plan includes establishing, by 2015, a national patent data centre, five regional and 47 local patent information centres (Murthy, 2011).

Another key innovation policy directive of the Chinese state is the goal of entering the top five countries for paper citation by 2015. The report, "The Statistical Data of Chinese S&T Papers", showed Chinese researchers published 1.14 million international sci-tech papers since

2003, ranking second in the world. These papers had a total citation of nearly 7.1 million times, ranking it fifth, moving up one place from 2012. The four countries with top paper citations are the United States, Germany, the United Kingdom and Japan (Chen, 2013).

In order to create a highly skilled national workforce within the next ten years, the central government released a new national development plan: the National Medium-and Long-term Talent Development Plan (2010–20). This plan is the first national comprehensive plan in China's history of national human resources development and is of vital importance to China's overarching developmental plan. It features concrete numbers of *rencai* (global talent) needed for specific sectors. For example, by 2020, more than 5 million *rencai* will be needed in equipment manufacturing, information technology, biotechnology, new materials, aeronautics and astronautics, oceanography, finance and accounting, international business, environmental protection, energy resources, agricultural technology, and modern traffic and transportation (Wang, 2010). The plan proposes that of every 10,000 people in the labour force, at least 43 professionals should be working on R&D, and R&D professionals will number 3.8 million by 2020. To put China's ambition in perspective, it is pertinent to look at the current number of R&D personnel in developed countries. According to Eurostat, there are a total of 1,356 million R&D personnel in the 27 countries of the European Union: including 284,300 in Germany, 211,100 in France and 175,500 in the United Kingdom. In contrast, the United States Department of Labor Statistics Bureau has put this number in the US at 621,700 (Wang, 2010).

FDI investment is another key component of the Chinese government's innovation development plan. China attracts the third-largest amount of FDI in the world, behind the United States and the United Kingdom. During the past five years, foreign companies have established hundreds of new R&D centres in China. According to several recent surveys, executives from multinational companies rated China as the most attractive country for future R&D investments. China has become a large exporter of high-technology products, accounting for one-fourth of China's total exports in 2005. What is important to emphasise here is that China's dependence on FDI as a driver of growth has, in fact, been carefully managed (Bach et al., 2006:508–10).

The management and control of FDI in China is clearly evident in the way the Chinese state classifies industrial sectors according to whether or not FDI is to be "encouraged", "restricted" or "prohibited". Indeed, a central component of the most recent Five-Year Plan (2011–16) is to shift away from FDI in labour-extensive manufacturing sectors. Hence, the plan places restrictive measures on foreign investment in traditional industries: "including the labour-intensive, polluting, energy-consuming or low-tech manufacturing and processing industries" (Opinions, 2010). Gradually, "China intends to guide the expansion of capability in some of these industries to its less-developed central and western regions. Investment in technology-intensive projects will continue to be encouraged" (Opinions, 2010). Obviously, the plan's strategic goal here is to ensure that all FDI is redirected into high-technology sectors (Opinions, 2010).

Fiscal policy is another important tool in implementing China's new long-term plan. The provision of tax incentives – perhaps the most novel policy – is designed to encourage company R&D investments. Suggestions include making R&D expenditure 150% tax deductible, effectively constituting a net subsidy, as well as introducing accelerated depreciation for R&D equipment worth up to 300,000 RMB (Schwaag Serger and Breidne, 2007).

Public procurement is also an important new instrument for promoting innovation in Chinese companies (Schwaag Serger and Breidne, 2007). This policy asserts that Chinese government agencies and entities must purchase domestic goods, works or services except where those goods, works or services can't be obtained within China under reasonable commercial terms (Schwaag Serger and Breidne, 2007). To qualify for inclusion in these catalogues the product not only had to be made in China, the intellectual property on which it was based also needed to be Chinese or transferred to China. The new policy was to make procurement open to foreign-owned companies in China as long as the innovation occurred on Chinese soil and they moved the R&D to China. However, after intensive foreign lobbying, Chinese agencies took steps to officially rescind the policy in July 2011 (Atkinson, 2012).

Furthermore, in a bid to further strengthen China's industrial base and move from a factor-driven economy to investment-innovation-driven economy, China has developed a number of policy directives designed to develop innovative clusters and national economic and technological development zones (ETDZs). To date, the Chinese government has successfully incubated entire industry sectors such as solar and wind power, battery technology and electric vehicles. Indeed, according to an OECD report, China has excelled at mobilising resources for science and technology on an unprecedented scale and at exceptional speed (OECD, 2011).

Moreover, China has been intent on using its huge market as an asset to develop distinctive standards with an expectation that its standards will be taken to an international level, in ways that small countries are not able to do. Lester Ross has asserted that market size and conditions "where dynamic technological developments threatened to eclipse existing standards" are the factors which encourage Chinese policy-makers to formulate domestic standards "in the expectation that market size may result in international adoption of China's standard" (Yan, 2007). The overarching thrust of this policy position is on the development of market power as a substitute for technological weakness. This approach is expounded in China's "indigenous innovation" policy. The idea of indigenous innovation endeavours to facilitate domestic technology breakthroughs rather than an overt reliance on the borrowing of imported technology from other countries. Conversely, a domestic standard for China's 3G network, it was conjectured, would operate to provide Chinese companies with a comparative advantage that would provide them with an opportunity to gain market share in a sector dominated by European and US suppliers. Hence, it can be seen that the rising interest of standard setting by the Chinese government is actually a strategic response to globalisation and the global economy, where standards have become an important tool to leverage gains in international production networks (Yan, 2007).

Thus, it can be argued that via a mix of bureaucratic and fiscal state capacity the Chinese state has recently devised a state-led "grand innovation" development strategy designed to facilitate the innovative upscaling of the Chinese nation state and the development of core indigenous-owned, high-technology intellectual property. The focus of these policy directives, it has been argued, contains a distinctively techno-nationalistic thrust. However, it can be argued that the Chinese state's strategic attempt to develop indigenous technology is a key component of its globalisation and alliance capitalist plan. This is because without their own intellectual property rights and assets Chinese firms will lack any bargaining power or intellectual capital from which to play an active role in the construction of, or gain access to, contemporary socio-technological ecosystems.

Furthermore, as I will highlight in the next section, the Chinese innovation developmental plan also contains a number of strategic policy directives designed to facilitate globalised relational capacity. What is important to emphasise at this point is that, in the global knowledge economy, key policy attributes that serve to frame the construction of successful development plans are bargaining power, resource and cluster development, coordination mechanisms and the development of regulatory capacities. This is because the key role of the state in this environment is to incorporate new social groups, associates and organisations into bargaining processes and to find the right balance between various particularistic interests. This requires the development of collaborative architectural planning and coordination mechanisms.

The collaborative economy, alliance capitalism and the Chinese government

The advent of the collaborative economy and the need for continuous systematic innovation at a global level has created new sets of system pressures and economic opportunities for economic actors and organisations. For Chinese policymakers, the nature and type of strategic development choices and capacities required in a context of collaborative global innovation are very different from those required when the policy objectives were solely centred around the idea of indigenous technology and the idea of building national champions. Indeed, systems complexity, alliance capitalism and collaborative organisation require a fundamental shift in the way governmental policy interacts with the emerging global technological system.

Firstly, as I established in the previous chapter, the state's territorial jurisdiction over knowledge is becoming less important than its capacity to access and organise complex networks of collaborative and dispersed innovation processes. Certainly, the fact that technological innovation and product development in the contemporary knowledge economy occurs in globally and organisationally dispersed innovation networks has significant implications for Chinese technology developmental policy and governance structures. Indeed, it has been argued that because these technologies are created and developed in contemporary innovation networks that are increasingly defined by economic space instead of political geography they transcend national or geographical governance structures.

Hence, it is very important that the state understands the deep structure and manoeuvring spaces of dispersed Collaborative Global Innovation Networks (CGINs). At a systems level, these CGINs are defined by self-organisation via collaborative alliances and dynamic adaptability. The shaping of these systems will occur in a globalised adaptive ecology of system actors committed to the construction of global collaborative platforms for the co-development of its technological assets and markets and the transnational sharing of appropriated value. Furthermore, as it has already been established, the power to shape or structure the overarching architecture of a technological system is defined by distributed agency. This means that the ability to influence the socio-technological systems development is intimately connected to an actor's ability to collaborate with technology producers, users, evaluators and regulators in developing technological ecosystems' specific design heuristics, testing standards, regulatory schemes and commercialisation processes.

It is precisely these changes in the global socio-technological system that have spurred the evolution of new forms of developmental states such as the GNS. Specifically, the need to develop new sets of developmental capabilities that address the existence of critical ecosystem dependencies and the relational capacity to participate in and form networks and their associated technological ecosystems are important strategic capacities required by nations wishing to engage in high-technology innovation at a global level. The specific sets of policy instruments and mechanisms that allow the GNS to achieve its key goals include the ability to source and provide information about transnational corporations and world markets to domestic firms and the development of institutional frameworks and funding decisions designed to provide the necessary conditions for stimulating connections between domestic and international companies. That is, the GNS's overarching goal is to facilitate collaboration and encourage innovation at both domestic and global levels (O'Riain, 2000).

Indeed, it can be argued that trans-territorial state capacity and ecosystem-shaping capabilities are important new state-policy tools that contemporary developmental states need to develop if they wish to construct an innovation-driven economy. The state can play a key focal role in the coordination of the collaborative process here as a network organiser and system integrator. By helping to facilitate the insertion of its domestic firms into a global innovation ecosystem, and by integrating international firms into its domestic economy and innovation platforms, the state can spur the development of a complex network of extra-territorial linkages that facilitate the development of globalised technological niches and ensure the interoperability and openness of key emerging innovation platforms and their associated technologies. In this way, governments can ensure the technological advancement of their domestic firms as well as the ability to appropriate technological rents from multiple globalised sources.

Hence, an important question that needs to be asked at this point is: to what extent does the Chinese developmental state have the capacity to actively create and shape new market for technologies by contributing to the production and co-shaping of socio-technological sectors? Does it have the state capacity and relational legitimacy to actively recruit the right network of private and public actors required to construct and sustain technological ecosystems and market structures at both domestic and global levels?

It can be argued, that the Chinese developmental state's relational capacity has increased significantly since the implementation of its innovation development policy in 2006. Certainly recent policy directives indicate the Chinese state has undergone significant policy adaptations, developed a dense network of global technological and market connections, and devised new forms of innovation capabilities that are designed to allow it to develop connections with a broad range of foreign counterparts as well as with transnational regulatory bodies. Indeed, it can be argued that the Chinese government has responded to recent developmental constraints by acquiring and projecting influence within global market governance and high-technology development areas as a "fledgling regulatory state", market "rule maker" and "technological network organiser". For instance, China has endeavoured to gain international recognition for its home-grown technology standards through regulators such as the ISO, while at the same time seeking to strengthen its negotiating power vis-à-vis the WTO by identifying standards dependency as a potential national security risk (Strange, 2011).

Primarily, both network organisers and system integrators' key role in the contemporary global socio-technological system is to focus on developing long-term reciprocal trust and forward-linked incentive structures to induce system actors to join and participate in the socio-technological network structures that will generate the most value and socio-technological development for their nation. Indeed, it has been asserted that: "In a world of growing specialisation and reliance on other companies for key elements of business value, the ability to build trust quickly and effectively represents a significant source of strategic differentiation" (Hagel and Brown, 2005).

These concepts provide important cognitive frameworks for policymakers who need to adjust their focus from an overt focus on national intellectual property appropriation and ownership towards the idea of dispersed innovation and the need for long-term, trust-based extraterritorial relationships. In this system, where the creation of value is globalised and the development of innovative assets is collaborative, states need to allow foreign actors to assume part of the value creation, and appropriation of parts, of any collaborative asset that is made in the socio-technological ecosystem. The notion of a reciprocal long-term, trust-based network differs substantially from conventional approaches to trust, which primarily looks backward (Hagel and Brown, 2005). That is, actors are specifically focused on whether or not the other party has delivered against expectations in the past. This approach is best practiced in static environments. This is because the past is a reasonable predictor of the future, and the time required to establish this type of trust is less critical. In the contemporary socio-technological system, with its rapid change and highly connected globalised systems environment, the focus has shifted to forward-looking incentives as the primary mechanisms to build trust quickly. This is because in highly dynamic, fast-paced environments the past is less and less useful as a predictor of future performance (Hagel and Brown, 2005).

Instead, what matters to system actors and coordinators is a clear evaluation of current capacity and the potential to accelerate capacity building so that participants can adapt successfully to these rapidly changing environments. Hence, it has been asserted that dynamic market environments require a more dynamic view of trust, one that is focused on the "shaping" of future capacity building as a way to accelerate trust building in the near term (Hagel and Brown, 2005).

From this theoretical perspective, the skill of partners is only a small factor in the development of trust over time. What it is argued to be more important is that a participant's current skill set is its "will" to acquire more skills and broaden its system capabilities, as Hagel and Brown (2005) assert. "Will can trump skill because partners, with proper motivation, will invest aggressively in building the necessarily capacities" (Hagel and Brown, 2005).

> "Will" from this theoretical perspective is a function of incentive systems – are the parties sufficiently motivated to deliver the promised outcomes? The will to perform can be shaped by both positive (reward) and negative incentives (penalties). Research indicates that rewards generate a more powerful and enduring form of motivation. In this environment the provision of cash rewards is not the central motivation, instead the focus is on helping firms accelerate their capability building and the building of long-term relationships and the development of multiple capabilities in order to facilitate longtime participation in network configurations. This requires the provision of frameworks for joint knowledge building. (Hagel and Brown, 2005)

Hence, the goal of successful high-technology governance in a contemporary context is not to control the socio-technological environment and appropriate nationally controlled intellectual property assets or to focus on the protection and development of nationally developed sectors. Rather it is to enable and adapt, and partially shape via collaborative coordination, the emerging socio-technological environment. Conversely, the ability to self-organise, adapt and develop extra-territorial generative relationships are all important system capacities required by the current socio-technological system.

Recent policy directives issued by the Chinese government do indeed indicate that the Chinese developmental state is currently engaged in a strategic attempt to generate extra-territorial relationships and a desire to be involved in the co-development and shaping of contemporary socio-technological ecosystems in a more open collaborative mode of development. For example, the national development program of strategic emerging industries during the 12th Five-Year Plan (FYP) period adopted by the Chinese government calls for closer international exchanges and cooperation, and a path of open innovation and internationalised development. Specifically, the newest FYP in article XVIII specifies the need to strengthen scientific and technological openness:

> Actively carry out all-round, multi-level, high level scientific international cooperation, to strengthen the scientific and technological exchanges and cooperation between the Mainland and Hong Kong, Macao and Taiwan regions. Increase the intensity of the introduction of the international scientific and technological resources, and participation in international science programs and big science projects around the country strategic needs. Encourage our scientists to initiate and organise international cooperation in science and technology plan, initiate or participate in international standard-setting. Strengthen the introduction of technology and cooperation, encourage enterprises to carry out the shares of M&A, joint research and development, patent cross-licensing and other aspects of international cooperation, support enterprises and research institutions to establish overseas R&D institutions. Open and cooperative efforts to increase the national science and technology programs, support of international academic institutions, multinational corporations to set up R&D institutions in China, to build domestic and foreign universities, research institutions joint research platform to attract global technological talent to innovation and entrepreneurship in China. Strengthen non-governmental scientific and technological exchanges and cooperation. (CPC Central Committee and the State Council, 2012)

Recently, the Ministry of Science and Technology lifted the limits on foreign investment in research areas to encourage more overseas companies to build technical centres in China. For example, the most recent foreign investment catalogue, published on 13 April 2010, is defined by its liberal attitude towards foreign investment and its recognition that R&D needs to be undertaken via collaboration with foreign firms. On 6 April 2010, the State Council issued "Several Opinions of the State Council on Further Improving Foreign Investment Utilization Work" (Guofa No. 9, "Opinions"). The State Council "Opinions" states:

> foreign investment in developing high and new technologies will be encouraged. China will continue to improve its hi-tech enterprise recognition system to enable foreign-invested enterprises benefit from the status and provide supports to the qualified Sino-foreign joint technology development projects. Multinational companies are encouraged to set up in China their functional centres such as regional headcounters, research and development, procurement, finance and settlement centres. In addition, The Opinions provide that foreign investors are encouraged to participate in reorganisation of domestic enterprises by means of takeover, equity acquisition etc. China encourages foreign investors to become strategic investors of the companies listed on its domestic stock exchanges, and will continue to reinforce regulation on foreign investment in domestic securities and on acquisition of domestic-listed companies. The Opinions also provide that China will expand the qualifications for foreign issuers authorised to issue RMB-denominated bonds to allow more issuers to benefit from domestic financial resources. The Opinions confirm that China is looking to having more qualified foreign-invested enterprises listed on its domestic stock exchanges. The Opinions also state that China will expedite, on the pilot basis, the process of market opening to foreign-invested guarantee companies. It will continue to encourage investment from foreign-invested venture capital and private equity funds, and to improve the regime for their exit. (CPC Central Committee and the State Council, 2012)

Furthermore, the most recent policy documents issued by the Chinese government are also focused on accelerating the pace of innovation and creating an open innovation system in which competitive pressures encourage Chinese firms to engage in product and process innovation not only through their own research and development but also by participating in global research and development networks (CPC Central Committee and the State Council, 2012).

In conclusion, it can be argued, that the recent spatial reorganisation of the global economy and advent of a more open collaborative model of technological development provides an important developmental window for the Chinese developmental state and innovative Chinese high-technology firms. By exploiting the increasing collaborative nature of the global socio-technological system the Chinese state is beginning to build up a critical network of interdependent alliance partners that are focused on achieving technological "convergence" and "interoperability" across the ecosystem platform. It is also expanding the technological governance arena in which it operates from one focused solely on achieving national development goals to one that is focused on achieving ecosystem development goals.

In the following chapter, I will examine exactly how the Chinese developmental state is employing a globalised, state-led, alliance-based developmental approach in the global wireless communication sector. I will highlight how this new strategic developmental strategy is defined by an attempt to combine the development of core indigenous technological assets with the development of collaborative R&D centres and socio-technological alliances with foreign firms in order to facilitate the development of collaborative dependencies and global network embeddedness in the sector.

5
Beyond Neo-Techno-Nationalism: An Introduction to China's Emergent Third Way: Globalised Adaptive Ecology, Emergent Capabilities and Policy Instruments

In order to respond to the rapidly changing and increasingly complex global socio-technological environment outlined in the preceding chapter and to participate in the development of the 4G technological ecosystem and gain core technology and market share, the Chinese government is currently engaged in an extensive overhaul of its technological strategy. It is replacing its go-it-alone techno-nationalistic strategy with a globalised adaptive-policy framework that is focused on the pursuit of technological alliances and global ecosystem embeddedness. It is important to note here these changes are not just confined to the global wireless communication sector, but also can be seen in other technological sectors such as aviation, nanotechnology and automobile technology (Tang, 2011; Ting and Shapira, 2010, 2011; Boeing, 2012; Quan, Haifeng and Zhenhua, 2014; Goedeking, 2010; Wang, 2008; McKinsey, 2013). However, as it has already been established, the focus of this book is confined to the global wireless communication sector.

Beyond the earlier FDI and techno-nationalist strategies, this new development strategy can be conceptualised as the third major attempt by the Chinese government to employ an effective industrial development strategy. As this chapter will highlight, the primary focus of this new development strategy is its attempt to combine the development of core indigenous technological assets with the development of collaborative R&D centres and socio-technological alliances with foreign firms in order to facilitate the development of collaborative dependencies and global network embeddedness. Its key strength is the way it responds to contemporary socio-technological systems change in a bid to generate the relational ties and collaborative alliances that are essential for the development and appropriation of technological knowledge assets and commercialisation processes. This is essentially a "third way" whereby the state's strategic control over the globalisation process is modified in a way that still allows it a certain degree of regulatory control and strategic agency, but in which the interests of global firms are actually integrated and embedded into the Chinese developmental strategy.

The capacity of the Chinese state to achieve its new developmental program is heavily reliant on its capacity to use its large market as a bargaining chip to induce foreign firms to form alliances with Chinese firms and conduct collaborative R&D. In order to motivate foreign firms and technological actors to engage in alliances with Chinese firms, this strategic plan includes policy directives to upgrade the Chinese intellectual property system and its enforcement procedures. Without a fair and comprehensive intellectual property system that also addresses system changes in intellectual property rights and appropriation methods, the Chinese developmental state will not be able to protect either the intellectual property of its own newly emerging innovative firms or the intellectual property of foreign firms, who will then be disincentivised from participating in technological alliances in any substantial way with Chinese firms.

It is also highly dependent on the capacity of the Chinese government to extend its developmental capacity beyond the territorial boundaries of the Chinese state into the emerging technological ecosystems and knowledge networks that define the contemporary technological system. This requires the Chinese state to become a key ecosystem constructor and shaper that can facilitate innovation in key technological sectors and motivate others to join these ecosystems and co-develop the high-technology products and intellectual assets necessary for the effective functioning of the ecosystem. Indeed, network organisers as strategic system actors play a fundamental role in shaping the evolution of their respective high-technology sectors.

In this chapter the institutional, regulatory and bargaining strategies that the Chinese government is using to gain the global leverage and legitimacy that will allow it to engage in global alliance ecosystem building and cooperative R&D as a development strategy in the global wireless communication sector will be examined. Specifically, I will highlight how the Chinese government is attempting to ensure the development of global alliances between foreign and domestic firms by building the indigenous capacity of its domestic firms in order to build up a critical mass of key patents and globalised organisational capacities that can be used as currency to facilitate technological alliances for the co-development of high-technology assets between its domestic and foreign firms. It is also attempting to ensure that foreign firms engage in technological alliances with Chinese domestic firms via its creation of a comprehensive funding and incentive program designed to attract foreign firms to set up R&D centres in China.

This new developmental strategy acknowledges that in the contemporary socio-technological system there exist sets of critical ecosystem dependencies that need to be addressed by all ecosystem actors before a technology can go to market. It also acknowledges the fact that in order to address these critical ecosystem dependencies, the Chinese state as a strategic developmental facilitator needs to develop its relational legitimacy at both a global and a trans-territorial level. This shift in the developmental objectives of the Chinese developmental state can be seen manifest in ways in which the Chinese government has undertaken active coordination and negotiation with major stakeholders in the 4G Time-Division Long-Term Evolution (TD-LTE) sector in order to expand its indigenously developed standard, TD-LTE globally, to promote collaborative breakthroughs in key technologies and increase the membership of the industry supply chain. Indeed, as this chapter will highlight, the Chinese government has played a fundamental role in both the global construction of trans-territorial alliances and the coordination of the TD-LTE full value chain itself.

It is important to note here that the avocation of joint cooperation between domestic and foreign companies and the sharing of key technologies are all policy directives that were expounded on in earlier policy areas, such as "trading market access for technology" in the era of FDI. Hence, they are not new in themselves. What is new is the incentive structures being employed to attract foreign participation and the underlying policy objectives and market-building goals. Whereas the goal before was to gain the necessary high-technology knowledge required to launch Chinese national champions into the global market, the goal now is the construction of hybrid firms that are both Chinese and foreign in ownership structure via alliance mechanisms that allow Chinese state leaders to appropriate value from network and ecosystem collaboration, coordination, and embeddedness, rather than from industry or sectoral monopolisation.

This behaviour is an important example of a state-led adaptive response to the emergence of rapid-innovation-based technological ecosystems and the emergence of new technological markets and appropriation mechanisms. That is, it is an adaptive strategy designed to adjust to technological systems change. Specifically, it represents a new round of capability building for the interventionist Chinese developmental state, in that new policy tools and methods of strategic global economic diplomacy are being devised and experimented with.

This chapter will be divided into two sections. The first section will offer a reprise of FDI and techno-nationalism as technological development strategy in order to establish a clear understanding of why these strategies need to be upgraded and adjusted in order for the Chinese state to achieve the high-technology status and economic growth beyond the middle-income trap. In the second section, I will develop an account of China's new alliance-based, state-led approach with specific emphasis on the role of the Chinese state as a strategic technological and economic coordinator and facilitator in the global wireless communication sector.

Foreign domestic investment and shallow industrialisation

As it has already been established in earlier chapters, China's first approach to technological development focused on attracting FDI to leapfrog the economy. This was a common strategy employed by developing countries with minimal capital reserves. It allowed for the importation of capital, equipment and technology to help build an industrial base (Thun, 2006:3). Since the onset of the reform process, China has attracted hundreds of billions of dollars in FDI and trillions of dollars in non-public investment (Zheng, 2005:19).

The impact of FDI on economic growth is one of the most controversial topics in the international political economy. The modernisation hypothesis asserts that FDI promotes economic growth by providing external capital, and through growth, it spreads the benefits throughout the economy (Thun, 2006:13). It is the presence rather than the origin of investment that is considered important because FDI brings with it advanced technology and better management and organisation. In the modernisation hypothesis, FDI is seen as a critical engine for economic growth in developing countries (Thun, 2006:13).

By contrast, the dependency hypothesis argues that whilst possible short-term positive impacts on economic growth result from FDI flows, there are often negative long-term impacts, as reflected in the negative correlation between the stock of FDI and growth rates. In the short run, an increase in FDI enables higher investment and consumption and thus contributes to economic growth (Thun, 2006:13). However, as FDI accumulates and foreign projects take hold, there will be adverse effects on the rest of the economy that reduce economic growth. This is due to the intervening mechanisms of dependency, in particular, "decapitalisation" and "disarticulation" (Thun, 2006:13).

Certainly, a plethora of empirical evidence reveals that contrary to its significant contribution to China's economic development, FDI has in the past played a limited role in transferring technology to key Chinese technological sectors and contributing to local technology capacity building. It has been estimated that in this period the average level of technology transfer via FDI was approximately two years more advanced than the existing Chinese technology base, whilst the technology gap between China and investing countries was estimated to be about 20 years (Young and Lan, 1997:12).

Conversely, whilst the Chinese party elite all share the conceptualisation that China must reform and open up its centrally planned economy, they are divided on the exact pace and content that reform measures should take (Bell and Feng, 2007:56). Given such results, the New Left, who are essentially techno-nationalists or mercantilists, are critical of neoliberal policy in its various guises whether as a rubric for free-market economics or as a broader metaphor for Western interpretations of modernity (Fewsmith, 2007:1).

The recognition that FDI was failing to facilitate technological upscaling of the Chinese nation led to a shift in policy direction by the Chinese government. Operating from the assertion that China's huge market of 1.3 billion people provides an important point of political and economic leverage, a techno-nationalistic strategy was devised that sought to develop China's own de facto technical standards (Bell and Feng, 2007:59).

Chinese strategic policy and the case of the wireless communication sector

In response to the New Left critique and the increasing recognition that the strategy of FDI has essentially failed, early attempts by China to reframe its development strategy focused on the development of its go-it-alone techno-nationalistic strategy. Here, Chinese policymakers attempted to leverage the size of the Chinese market as a bargaining tool to establish the perceived lock-in effects associated with technological standards ownership and the comparative advantages that it confers.

It was within this period the idea of indigenous standard development was explored and incorporated into policy documents. The primary emphasis of this policy program was the creation and commercialisation of proprietary ideas, standards and technologies created by Chinese companies. The development of the Chinese 3G standard, Time Division–Synchronous Code Multiple Access (TD-SCDMA) and Wireless LAN Authentication and Privacy Infrastructure (WAPI) fall within this branch of the Chinese government's standards and indigenous innovation technology policy framework.

In 2006, in accordance with the nation's new "indigenous innovation" policy, the Chinese home-grown technology standard, TD-SCDMA assumed the status of a flagship standard and acquired support from the highest level of the Chinese government (Gao and Liu, 2012).

The Chinese government offered four types of support for the standards development. Firstly, support through signalling. For example, strong support for the TD-SCDMA Industry Alliance (an alliance set up by Datang, the founding company to attract domestic and foreign collaborators) by the National Development and Reform Commission (NDRC), the most powerful government agency in China, indicated that the Chinese government wanted to support TD-SCDMA. This policy signal was very important in attracting other firms such as ZTE and Huawei, who are competitors of Datang, to join this alliance and increase the credibility of TD-SCDMA in attracting the interests of foreign firms (Gao and Liu, 2012).

Secondly, financial support provided by the Chinese government. For example, government agencies such as NDRC, Ministry for Industry and Information (MII) and Ministry of Science and Technology (MOST) provided 700 million yuan to facilitate collaboration between member firms of the TD-SCDMA Industry Alliance. Specifically, part of this money was allocated to member firms to pay Datang for sharing its TD-SCDMA-related technologies. This not only lowered the barriers for member firms to develop TD-SCDMA-based technology and products but also increased member firms' confidence in relation to government support and the future of TD-SCDMA (Gao and Liu, 2012).

Thirdly, strategic support provided by the Chinese government via the provision of technical service. For example, the former MII organised the MTNet test to verify the capability of the TD-SCDMA system to be deployed as a standalone network in 2004, rather than as a complement to Wideband Code Division Multiple Access (WCDMA) (Gao and Liu, 2012).

Fourthly, support provided through administrative order. For example, in 2006, the government, including NDRC, initiated the large-scale

TD-SCDMA Network Application Trial project and asked the telecom service providers to support TD-SCDMA trials in five Chinese cities. In 2007, the telecom service providers were asked again to support TD-SCDMA pre-commercialisation trials in ten cities, including Beijing and Shanghai. In April 2008, China Mobile was asked to offer TD-SCDMA service based on its pre-commercialisation network in Beijing during the Olympic Games. Conversely, as a direct result of the co-evolution process between Datang and the government, the gradual development of the TD-SCDMA value chain was able to occur (Gao and Liu, 2012).

However, like WAPI, TD-SCDMA failed to meet the techno-nationalistic aspirations of the Chinese government. Specifically, it failed to commercialise due to the fact that critical ecosystem dependencies were not addressed and compatible handsets and products were not developed due to the perception by domestic and foreign firms that the standards technology was inferior when compared to existing standards such as Code Division Multiple Access (CDMA).

Competition, compromise and alliance capitalism

In an important example of adaptive-state flexibility and capacity building, the Chinese government has recently yielded to international pressure and the changing nature of the global technological environment, with its very high skill intensity, a rapidly moving technology frontier, and technological obsolescence, and an incremental fashion again sought to revise its technological development strategy. This time supplementing its go-it-alone techno-nationalistic strategy with a policy framework that is receptive to a more open-market focus and which is designed to address the existence of critical socio-technological ecosystem dependencies in the global wireless communication sector as well as the need to meet the requirements of the international regulatory environment.

Beyond the earlier FDI and techno-nationalist strategies, this new development strategy can be seen as the third major attempt to mount an effective industrial strategy in China. It is also the first to achieve both significant growth in high-technology sectors and real cooperation between domestic and foreign firms. Indeed, recent efforts by the Chinese state to develop its own wireless technical standard and engage in pre-competitive R&D in order to maximise the relative gains that the nation receives from participation in the global economy, represents a new and exciting example of recent attempts by the Chinese government to develop its own competitive technology policy.

A contemporary example of China's attempt to reinvent its standards policy into a theoretically sophisticated strategic technology policy that has moved from the conceptualisation of indigenous standard development to pre-competitive architectural construction and anticipatory standards can be drawn from newly released documents detailing China's current strategic policy in relation to 4G wireless Internet development (Cajing, 2009). These documents indicate that the Chinese state has undergone a shift in its political agenda, technological focus and R&D expenditure allocations in relation to the development of wireless technology in the 4G sector. The emergent policy position is designed to facilitate the development of standards that both anticipate and operate to define the future regulatory global infrastructure and market governance structure as well as pursuing the development of network and research alliances with key stakeholders and actors who are central in emerging wireless technological ecosystems.

The Chinese state and the wireless communication sector

The information communication sector is a "crown jewel" in China's network of state-owned enterprises. China is the world's largest telecommunications market, with over 1.132 billion mobile subscribers and close to 564 million Internet users as of year-end 2012 (China Internet Network Information Center, 2013). It is also the world's largest producer of handsets. In mid-2008, China became the largest broadband market in the world when it surpassed the US. At the beginning of 2012, China's broadband subscriber population base had reached 380 million (Muncaster, 2012). With a penetration rate of 39.9% (China Internet Network Information Center, 2013), the Chinese domestic market has significant market potential. This contrasts with the more advanced markets, which have been saturated and are driven by replacement demand, whereas China's market growth is driven by original demand (i.e., there is still significant potential for future market growth). Because of China's low penetration base, the growth and future of the global wireless communication sector is intertwined with the Chinese customer base and hence attracts the interest of foreign companies.

It is this large market and customer base that affords the Chinese state a strategic bargaining tool that it has utilised to induce foreign firms to engage in collaborative alliances and technology sharing in exchange for market access. For example, research indicates that the bargaining power of MNCs is directly derived from their ownership or access to sophisticated technology, product differentiation (including strong brand names), their ability to contribute to exports (especially through intra-firm transactions), their access to or lower cost of capital and their product diversity (Low, Johnston and Wang, 2007). In order to establish a strategic position in the Chinese market, global companies have progressed from initial arm's-length export transactions to collaborative R&D, production and marketing relationships. Empirical evidence indicates that China's central government prefers to develop economic alliances with global, publicly listed companies, especially in joint ventures and alliances with local Chinese partners. For example, global companies, such as Motorola, Siemens and Lucent – who possess technological knowledge, product development, marketing experience and a prior history of economic legitimacy – are considered by China as attractive partner firms (Low et al., 2007). Motorola, Lucent and Nortel have established technical and research alliances with indigenous Chinese firms in an attempt to acquire technological legitimacy. Hence, they are actively collaborating with local partners (equipment manufacturers), customers (carriers), and the central government, and helping them participate in global technical and research alliance programs. As a result of this behaviour, legitimacy justification requirements espoused by network constituents are being met (Low et al., 2007), and it is these telecommunication companies that are assuming key network positions in the Chinese wireless communication sector. These include recognition as valuable corporate citizens; potential access to privileged information, including timing and awarding of licenses; preferences and incentives in setting up local research and development alliances; and the awarding of local manufacturing licenses. For example, foreign manufacturers including Nokia, Motorola and Qualcomm have become the direct and instant beneficiaries in the value chain of 3G businesses (Low et al., 2007).

What needs to be emphasised at this point is that the Chinese market forms an integral component of the internationalisation plans of major global companies. For instance, China was Ericsson's third-largest market in 2012: "A company cannot call itself a global leader unless it is also an industry leader in China, said Ma Zhihong, executive director and Asia-Pacific senior vice president of Ericsson" (Want China Times, 2013). The Chinese market comprised 7.6% of Alcatel-Lucent sales and 9.6% of Nokia Siemens sales (Want China Times, 2013).

The Chinese telecommunications sector is defined by the way in which all telecom operators are state-owned and are controlled by the central government. A 2010 government report of the central government enterprises ownership structure showed that government capital occupied 96.37% of the telecommunications industry at the end of June

2010 (SASAC, 2010). Conversely, foreign investors wishing to operate in the transitional Chinese economy, specifically its telecommunications sector, which is the subject of continuous structural reforms and elite control, have often been forced to operate within a range of settings, often subjected to different "rules of the game" (North, 2007 cited in Low et al., 2007:97). Such settings have been known to range from a highly marketised setting to one that is highly controlled and regulated, as is the case with state-owned enterprises (Low et al., 2007:97).

Furthermore, the Chinese government is frequently modifying the structure of the market itself in a bid to reorganise it, presumably in an attempt to find the optimal competitive market structure. One of the most significant changes in the structure of the Chinese telecommunications market occurred in 2008 when the market itself was transformed into three major full-service operators (Cajian, 2009). The restructuring merged six of the country's state-owned mobile phone and fixed-line operators into three nationwide carriers offering fixed-line and wireless services, instead of dividing coverage in terms of region or type of service (Cajian, 2008).

Indeed, the Chinese developmental state – at all levels – has played a leading, entrepreneurial role in facilitating innovation-led growth in the wireless communication sector. Certainly, the Chinese government considers the sector to be a strategically important industry that requires government steering, financing and regulation (Steinbock, 2003:5). However, in accordance with the key precepts pertaining to the flexible developmental state model (FDS), the government of China's policy-making response to the development of wireless technologies has been framed by a process of experimental trial-and-error policy learning and the implementation of policy directives, laws and regulatory mechanisms that seek to respond to the fast-paced developmental and commercial signals and changes that are occurring in the global wireless ecosystem. I have termed this approach "globalised adaptive ecology". This ecology is defined by the way in which the Chinese developmental state adjusts it policy directives in response to system changes and signals in order to acquire the system-shaping capabilities and relational legitimacy necessary for globalised technological ecosystem membership.

Hence, contrary to popular rhetoric, the government of China has not engaged in any predetermined developmental plan in the sector per se, but rather has sought to learn from the evolving socio-technological environment in order to gather information on barriers to market and innovation opportunities and to design specific policy sets to address these barriers as they exist at certain points in time (Liu, Quhong and

Ling, 2013). This does not mean the Chinese government does not have a developmental program for the wireless communication sector. It does, and this is clearly outlined in the 2006 and 2011 ten-year development plans. What it means is that these plans are subject to continuous evolution and co-evolve with changes in the socio-technological system itself. From this theoretical perspective, telecommunications governance can be seen as a highly dynamic, adaptive system (Cherry, 2007) that is continuously adjusting to system constraints and opportunities. As a direct result, the strategic policy directives issued by the Chinese government in the sector often appear confusing and seem to have conflicting objectives. However, a key desire that has been continuously articulated by the state and its representative ministries has been to ensure that Chinese domestic firms in the sector achieve global technological breakthroughs and market share. The Chinese government is not just interested in the development of technologies for the Chinese domestic market, but in facilitating the development of global technologies and ecosystem embeddedness. For example, in a personal interview, Xielin Liu, a professor and associate dean of the School of Management, University of Chinese Academy of Sciences (GUCAS), asserted that: "this is a critical issue for the Chinese government how to build/be involved in global innovation networks ... the government knows this is necessary but how to do this ... coordinate this ... is a major dilemma"(Liu, 2013).

4G: from neo-techno-nationalism to global ecosystem development

The Chinese government, operating from recent cues in the 4G TD-LTE socio-technological environment, has recently begun to develop specific policy measures and a global R&D strategy that will allow the Chinese government and its key domestic firms to play a fundamental role in the shaping of emerging 4G collaborative knowledge and commercialisation networks.

Former Chinese Vice Minister of MIIT Gou Zhongwen asserted:

The accelerated progress of technology makes the ability of businesses to utilise external knowledge increasingly ... Many MNCs are changing the strategy of treating competitors as enemies and are moving towards deep level technical cooperation and strategic alliance. ... The United States has set up 5000 new technology R&D leagues with foreign companies in key areas including new materials, information and biotechnology in the last 10 years. Before 1980, the revenue of the top 1000 companies in the United States from these technology leagues was only 1% of their total revenue. But by 2003, the figure had already reached 20%. Through cooperative R&D, MNCs not only can concentrate resources in tackling key technologies, but also can disperse R&D cost and investment risk, as well as benefiting from accelerated industrialisation and shortened the return period of technology investment through results sharing. This ability of utilizing external technologies makes it possible for MNCs to be highly flexible in response to competition in the international arena. (Gou, 2006)

Policy directives, press releases and company memorandums indicate that the Chinese government is highly responsive to changes in the 4G wireless sector. In 2007, the Chinese State Council announced a large-scale research plan for a 4G standard development. The research project, which was classified as a key national program, was entitled the "Next-Generation Broadband Wireless Mobile Communications Network" (Cajian, 2009). In response to the need to fund basic research in the sector a longer time frame of approximately 15 years was allocated for the research process. An initial investment for R&D of 20 billion yuan was given by the Ministry of Finance (MOF), with total investment in the project expected to exceed 70 billion yuan (Caijing, 2008). The bulk of this funding is for basic research: "The central financial administration will mostly invest in the early stages of basic research". Furthermore, "product development and industrialisation will mostly come from the market" (Caijing, 2008).

It is important to note here that the way the research and development process works in the wireless communication sector, specifically the need for technological actors to participate at an early stage of product development in order to ensure systems convergence and interoperability across the technological ecosystem, has raised fundamental implications for intellectual property rights and has led to the development of a new, more open-based collaborative model in the sector. Firstly, due to the need to engage in pre-competitive market research in order to address the existence of critical ecosystem dependencies and develop the required industry standards, competing companies engage in research on the same topics in order to present their solutions to the other standardisation partners. In this way, firms operating in the sector need to share their emerging technological solutions with their competitors (Brismark and Alfalahi, 2008). As a direct result, patent protection is a key prerequisite for any firm wishing to participate in the innovation ecosystem early-stage developmental processes. The key goal of this early-stage research process is to develop a set of standards that will define the interoperability of the system. When you have a standard, you have a specification; and the specification defines the market. This market is owned by all the companies that have contributed to the standard (Brismark and Alfalahi, 2008).

In traditional sectoral systems, a basic principle with patents is that when one company can obtain a monopoly it can stop others from using its patented solution. However, in the wireless technology sector the need for convergence and interoperability in standardisation has resulted in the development of a system whereby the partnering companies must sign a waiver from the beginning, by which they waive their rights to a monopoly and promise to license all of the involved patents on fair, reasonable and non-discriminatory (FRAND) terms. In telecom standardisation, the companies that contribute the most get the largest share. But an important part of this business model is that companies which take part in the development of solutions and products are guaranteed part of the returns instead of gaining a monopoly (Brismark and Alfalahi, 2008).

The way the telecom industry traditionally works with open standards can guarantee this, precisely because the sector selects core essential standards solely on their technical merits. Furthermore, a voting process decides the outcome, which ensures consensus among participating companies. The only way licensing can work in this developmental context is that all participants are open with what their technological developments are at an early stage in the research and standardisation process. The overarching goal here is to achieve interoperability via open standards. The open-standard process is unique compared to other industries in that competing companies continuously share their latest results, technologies and solutions (Brismark and Alfalahi, 2008).

The findings, patents and industry standards generated by this basic research and involvement in the standardisation process, it can be argued, are of significant importance to the Chinese developmental agenda. Firstly, the patents and intellectual capital generated from the research provide valuable currency from which the Chinese state and its key domestic firms can bargain with foreign firms in order to gain access to established technological ecosystems. As highlighted in Chapter 3, successful high-tech innovation and commercialisation requires partners that are both able and willing to participate in the development of complementary products and service solutions. Conversely, ecosystem membership is highly dependent on the fact that a firm possesses intellectual assets and skills that other ecosystem actors need to commercialise their own products (Adner, 2012). Alternatively, it provides the Chinese state with the intellectual capital necessary to construct its own technological ecosystem or niche positions and the power to select with which foreign firms it will form technological and market-sharing alliances. As has already been established, ecosystem alliances are required in complex technological systems due to the need to ensure interoperability and product convergence. For example, in the 4G TD-LTE sector, there needs to be compatible network equipment, infrastructure and handsets.

In order to address the need for early pre-competitive market research and the existence of critical ecosystem dependencies at an early stage in the development of the 4G LTE standard-making process, the Chinese government asserted at the beginning of the project that "unlike Datang's lone involvement in the development of the TD standard, China's homegrown 3G telecom standard, there will be multiple participants – Datang, ZTE and Huawei among likely candidates – in the 4G technology R&D" (Caijing, 2008). Moreover, the government also asserted that "the framework China has created for its 4G R&D is more open and advanced than that for the TD standard" (Caijing, 2008). On the new path toward 4G, China must take to heart the lessons gleaned from the TD process and build an open, market-based framework that welcomes all possibilities. As Hou Ziqiang, a researcher at the Institute of Acoustics (IOA) Chinese Academy of Sciences (CAS), told Chinese business magazine *Caijing*: "The [4G] future is settled. China and LTE will unite and move forward together. The chances of China proposing its own [4G standard] are slim" (Caijing, 2008).

Hence, in order to achieve both ecosystem embeddedness and ecosystem-shaping capabilities, policy documents released by the Chinese government and its key ministries indicate that the Chinese state is focused on: (1) ensuring its domestic firms develop the necessary core technologies that will allow them to compete in the global wireless ecosystem, and (2) facilitating the construction of the necessary institutional infrastructure and relational legitimacy that will provide important opportunities for it key domestic firms to engage in collaborative network building, joint R&D and global commercialisation processes. For example, in 2006, the minister for MIIT conjectured: "Competition among modern companies is to a big extent the competition over knowledge, technology and information. To obtain knowledge, technology and information, a company needs to rely on the R&D and innovation of its own on the one hand, and should also be good at utilizing the innovations of others on the other hand" (Gou, 2006).

Certainly, it is undeniable that recent S&T policy directives clearly articulate an intent to expand and develop future cooperation between foreign R&D firms and Chinese partners, and to improve prospects for further technology spillovers, with its emphasis on international cooperation and collaboration. It states that international S&T cooperation is an important aspect of China's policy of openness to the outside world: "China is ready to enter into cooperation with any country depending upon the needs of Chinese S&T and economic development, according to the principles [of] equality and mutual benefit, mutual enjoyment of benefits, protection of intellectual property rights, and respect for standard international practice. Cooperation can be bilateral, multilateral, with private parties or with governments in foreign countries at any level or through any channel" (Irwin Crookes, 2008). For example, Alcatel-Lucent and China Mobile recently signed a co-creation agreement to develop and test a series of lightRadio™ breakthrough technology elements. In August 2012, the companies extended this agreement by signing a Memorandum of Understanding (MoU) to conduct joint nanocell research in China. Nanocells are a type of radio access technology that integrates wireless small cells with carrier-grade Wireless Local Area Network (WLAN) access points (Communiqué de presse, 2012). Furthermore, Chinese delegates are increasingly populating international bodies' working groups (around 10% of the IEEE 802.16 members are from China, as are 20% of the "Potential Members", and very few have recently lost their membership). Moreover, a considerable number of Chinese companies are active in the two partnership projects and in the ITU30 (Graham, 2008).

Hence, as it has already been established in Chapter 4, the contemporary Chinese high-technological developmental plan contains two complementary strands. Firstly, there exists a drive to promote indigenous enterprise-level development to foster domestic intellectual property that can be licensed internationally and reduce the country's reliance on overseas technology inputs (State Council, 2006) as well as provide valuable intellectual currency to facilitate technological ecosystem membership. However, the fact that the plan also emphasises the need for international cooperation and collaboration indicates that the Chinese government is also currently engaged in the development of a delicate policy program that endeavours to combine the need for domestic technological and competitive upgrading with the need for international collaboration (Jakobson, 2008 cited in Irwin Crookes, 2008).

This new policy framework combines both indigenous and global policy initiatives in a way that is designed to enhance the ability of

Chinese domestic firms operating in the sector to actually shape the development of emerging 4G networks and achieve a high degree of ecosystem embeddedness. Indeed, it has been conjectured that:

> domestic firms are gradually experiencing the third wave of inno-
> vation named "indigenous innovation in open era" as to the great
> opportunity coming from global industrial restructuring, compared
> to the former two waves of innovation, which respectively are "indig-
> enous innovation in closed era" and "all-around introduction in open
> era". (Xiaolan and Xiong, 2011)

Certainly, key policy documents and personal interviews conducted for this book indicate that the Chinese government wants its domestic standard to have global, not just domestic, reach and that the Chinese government was aware that the immature nature of TD-SCDMA and the changing nature or the global wireless ecosystem meant that the path to 4G required a more open, international, alliance-based approach. For example, in a personal interview conducted at Tsinghua University in 2011, Professor Quhong asserted, "Chinese state leaders were becoming increasingly aware of the need for an internationalised approach in order to ensure the successful commercialisation and global adoption of its domestically developed indigenous standards and technologies" (Quhong, 2011).

In the following section, I will outline the key fiscal and organisational governance policy tools being utilised by the Chinese government to facilitate the technological upscaling of its domestic firms. These strategies are commensurate with the idea that the state as a developmental actor needs to play a key role in assisting its domestic firms to engage in cutting-edge R&D and enter global markets.

Go Global Policy

The Go Global Policy, initiated by the Chinese government in 1999, is an example of a state-led strategic policy program that is designed to encourage Chinese domestic firms to utilise external R&D resources and technological knowledge via introducing foreign technology or equipment, joint venturing, and collaborating with users, suppliers and public or private research institutions to facilitate the search for resource security and obtain higher returns for China's huge foreign exchange reserves, etc. (Cheung, 2009). State support plays a critical role for Chinese enterprises that want to go abroad. For example, Beijing provides subsidies and credits to Chinese companies attempting to enter key overseas markets involving energy projects and/or technology acquisitions. Chinese state-owned banks have expanded their overseas presence to help facilitate outward foreign direct investment (OFDI) as well as to increase investments in overseas finance markets. For instance, from 2007 to 2008, Chinese investment in foreign finance sectors increased seven-fold to approximately $14 billion. According to the Chinese Ministry of Commerce, this represented 25.1% of all of China's OFDI for the year (Ministry of Commerce, 2012).

Furthermore, in 2009, the Chinese government announced that it would be allocating a portion of its foreign reserves specifically to support the attempts of Chinese enterprises to move into foreign markets. Also, the China Investment Corporation (CIC), the sovereign wealth fund, has engaged in a targeted campaign to expand purchases of shares of foreign companies. Primarily, China's outward investment has a dual purpose of building China's political capital and influence around the world. China's chosen route to economic expansion has therefore been closely aligned with its strategy to strengthen its global political presence. Specifically, Chinese domestic firms and political leaders have worked in tandem to build strong relationships with developing countries. Companies like Huawei and ZTE have used their experience in building China's own markets to develop new ones in other emerging economies before tackling developed economies. Their better understanding of emerging markets provides a stronger guarantee of success in their initial overseas expansion plans, improving chances of a smoother entry into more developed Western markets later on. Meanwhile, China has been building alliances with other developing economies in political forums and multinational negotiations.

Despite recent media speculation of a slowdown in China, Chinese OFDI continues to grow. Indeed, in December 2012, China recorded its highest levels of Chinese OFDI; and in a first, OFDI exceeded inbound FDI. According to the Ministry of Commerce, Chinese OFDI for the year 2012 was US$77.2 billion, up from US$60.1 billion in 2011 (Ministry of Commerce, 2012).

Chinese government funding for domestic firms

State capitalism is defined by the way in which it provides funding to both state and privately held domestic firms in the form of credit, contracts and subsidies. These subsidies are provided by the state to help key domestic firms compete with foreign MNCs, in an attempt to help them capture a dominant role in both the domestic economy and export markets (McCarthy, 2012). For example, in the 12th Five-Year Plan approved by China in March 2011, next-generation information technology is identified as one of seven "strategic and emerging industries" for priority government support. It calls for increased governmental support through state-owned bank financing and credits at "preferential rates" from the Export-Import Bank of China. The policy also called for increases in export tax rebates and more use of export credit insurance by the industry (Accenture, 2012). The Chinese government is therefore playing the role of a private equity investor/venture capitalist by providing capital to help its domestic firms develop core technologies and acquire global technological ecosystem embeddedness. Unfortunately, the exact amounts and governmental origins of this type of support are often not reported or disclosed; hence, it is not possible to fully understand either the level or types of support being provided.

Nevertheless, in line with this state-led policy program, the state-owned China Mobile and the privately owned Huawei and ZTE have all recently been awarded a variety of forms of government support, including direct grants, preferential tax treatment and equity infusions in order to facilitate the development of their core technologies and global market expansion (McCarthy, 2012).

In 2010, Huawei reported receiving 433 million yuan in unconditional government grants and 545 million yuan in grants that were conditional on completing certain research and development projects. In 2009, Huawei reported receiving $129 million in government grants. ZTE received 471 million yuan in government grants, contract penalty income and other miscellaneous gains in 2010, according to its annual report. In 2009, ZTE reported receiving $92 million in government subsidies, including grants, support for technology development and tax subsidies. Neither company has disclosed the volume of government grants received in 2011 (Accenture, 2012).

In addition, telecom equipment manufacturers that qualify as new and high-technology enterprises are eligible for lower across-the-board tax rates. ZTE reports that numerous subsidiaries enjoy a 50% reduction in their income tax rates due to this status, while other subsidiaries have been granted temporary tax holidays based on this status or additional provincial and local tax incentives. China also refunds VAT taxes paid to companies in certain industries, including rebates on software procurement. ZTE reported receiving 1.9 billion yuan in such refunds and other tax subsidies in 2011. While Huawei does not disclose its Chinese tax rate or the eligibility of any of its operations for preferential treatment, its effective tax rate was 6.5% for its global operations in 2011, far below the statutory rate of 25% in China (Accenture, 2012).

Huawei and ZTE have also benefitted from direct equity infusions from China or have been supported by state-owned financial institutions. Huawei received an infusion of $5.8 billion from its equity holders in 2009. The company is 99% held by the union of its employees. There is very little information about the true ownership structure of Huawei and the nature of its employees' ownership of the company. However, in China, all unions must be part of the All China Federation of Trade Unions, a public entity associated with the Communist Party. The equity infusion was equal to nearly 4% of the company's sales revenue in 2009 (Graham, 2008).

In 2008, ZTE issued 40 billion yuan in bonds cum warrants, which were guaranteed by the state-owned China Development Bank. The bonds, which bear an interest rate of 0.8%, have permitted the company to fund major capital investments. In addition to being backed by major state-owned banks, it appears that many of the major purchasers of the bonds are themselves state-owned firms. The 40 billion yuan the company has been able to raise through bond insurance was thus directly supported by government guarantees and government purchases, resulting in a major government-backed infusion of funds to the company (McCarthy, 2012).

Finally, as a government-owned company, China Mobile's push for LTE has received firm support from the government. A new policy directive by the MIIT says 3G and 4G will be made as the "twin engine" for telecom growth during 2011–15, and total investment will exceed 1 trillion yuan (US$160 billion), half of which will be used for LTE. Approximately 60% will be for network equipment purchases, 30% for construction and 10% testing and other auxiliary products (McCarthy, 2012).

Huawei has asserted that both central and national government support has played a crucial role in the development of the company in its early stages. As Huawei President Ren Zhengfei asserts: "Huawei was somewhat naive to choose telecom-equipment as its business domain in the beginning. Huawei was not prepared for such an intensified competition when the company was just established. The rivals were internationally renowned companies with assets valued at tens of billions of dollars. If there had been no government policy to protect (nationally owned companies), Huawei would no longer exist" (Xiao, 2002:127).

This state-led approach to fostering high-technology development is currently a point of much contention in Europe and in the US, both of whom assert that it distorts the global market and provides Chinese domestic firms with an unfair trading advantage. For instance, in 2010, the EU initiated an investigation into subsidised imports of wireless wide-area networking modems from China after receiving a complaint from Option N.V., a Belgian producer of such wireless modems. The complaint primarily targeted Huawei and ZTE, and it stated that the Chinese exporters were able to flood the European market with low-priced products due to heavy subsidisation by the Chinese government. Following a preliminary investigation, public reports state that the EU was proposing significant duties of more than €30 for the imported Chinese modems, which normally only cost between €20 and €30, highlighting how the extent of subsidisation was in the triple digits. Prior to imposition of the duties, Option N.V. and Huawei entered into a "cooperative agreement" which included Huawei paying €33 million to license some of Option's software and Huawei purchasing Option's subsidiary, M4S, for €8 million. In the wake of this agreement, and, "in the spirit of future collaboration", Option then withdrew its complaints and the investigation was terminated (McCarthy, 2012).

Public procurement and trading technology for market access

Ernst (2002) and Ahrens (2010) argue that government procurement contributes to innovation by bridging the finance gap, focusing on market signalling, lowering the risk of research and development, and stimulating demand (Liu and Cheng, 2011). From 2006, China's indigenous innovation strategy has used technology public procurement as one of the main drivers for indigenous innovation, and Chinese domestic enterprises have taken on the task of incubating new technologies (Liu, 2013).

Public procurement can also help promote innovation and accelerate the diffusion of innovative products and services in the economy. The size of the Chinese market, its dynamism and the important roles played by the central government and sub-national authorities in the Chinese economy point to the strong potential for promoting innovation via public demand. The volume of government procurement has been expanding rapidly, although, at about 2% of GDP, it is still far below the levels in more developed countries. The Chinese government has recognised this potential and attempts to make use of it. The Medium to Long Term Plan (MLP) for the Development of Science and Technology (S&T) for the first time assigns public demand an important role in economic development and the promotion of innovation. This represents a policy innovation, since the Chinese government traditionally relied entirely on supply-side policies to promote technology development. The development and implementation of an innovation-oriented procurement policy is a demanding process in terms of the required expertise and the coordination of the government agencies involved. Innovation through public procurement cannot be "ordered"; rather, it has to be the result of a sophisticated articulation of demand for innovative products or services and of a transparent competitive process (OECD, 2007).

For example, Chinese Mobile has recently awarded initial 4G contracts totalling approximately 20 billion yuan ($US3.2 billion), with Chinese firms securing more than half of the contracts and foreign firms winning only about a third (Chinascope Financial, 2013). Shenzhen-based Huawei and ZTE have each obtained approximately a 25% stake out of the total 4G procurement in China Mobile's tender in this round. European vendors Ericsson, Alcatel-Lucent, and Nokia Siemens have each obtained a share of around 10% (Chinascope Financial, 2013).

Relational legitimacy versus forced technology transfers

However, whilst the preceding policy and fiscal tools are utilised by many governments to help their domestic firms develop high-technology innovation capabilities, the problem for Chinese policymakers is that the idea of "forced technology transfer" is consistently met with international action and subsequent failure. Furthermore, without relational legitimacy and transparency, membership in globalised technological ecosystems and access to foreign markets will be inhibited. Certainly, policy documents indicate that the Chinese bureaucratic infrastructure's top-down industrial policy-making style is becoming increasingly difficult to employ in an environment where the development of relational assets is a key prerequisite required to facilitate successful network development, and both domestic political associations and foreign multinationals exert a greater degree of influence over the decision-making process. For example, as Scott Kennedy articulates, today's firms influence policy directly through the lobbying of regulators and indirectly through business associations and other intermediaries. The success of these firms is based on their ownership type, size and technological sophistication (Kennedy, 2005).

For example, in April 2010, the Chinese government was again forced to retreat from an aggressive policy attempt to force foreign firms to transfer R&D via alliances with Chinese domestic firms. The "Buy

Chinese" policy directives issued by the Chinese government in April 2009 sought to force foreign firms to transfer its R&D platforms via co-development and joint patent registration. The "Buy Chinese" procurement policy directives, it was argued, will favour Chinese producers in six technology areas – computers, clean power, communication, office equipment, software and energy-efficient products. Companies were given just three weeks to apply to be treated as domestic suppliers, a status trade groups say few were likely to qualify for, even those such as General Electric and Microsoft that have research and development centres in China (Chao, 2010).

US companies appealed to former Secretary of State Hillary Rodham Clinton and Treasury Secretary Timothy Geithner to make the procurement plan a priority in their annual Strategic and Economic Dialogue with Beijing. The procurement policy would create barriers to competition in the Chinese market for the most innovative American companies, asserted a coalition of 19 American groups including the U.S. Chamber of Commerce and the Business Software Alliance: "some companies would consider pulling out of China if they conclude the loss in sales will be too great, U.S. and European trade groups say" (Chao, 2010). The government is China's biggest software buyer and a key customer for other technology. Losing that market might hurt companies including Microsoft Corp., Intel Corp. and Motorola Inc. Suppliers worry the rules could be extended to purchasing by major state-owned companies in power, telecom and other fields. Furthermore, a separate coalition of 34 groups for technology companies and manufacturers in Europe, Japan, South Korea and Canada also has appealed to Beijing to reconsider. In April 2010, fearing protectionist reactions from the Obama administration, EU, and foreign companies, China modified the plan to address the concerns of foreign firms and governments (Chao, 2010).

What this outcome indicates is that the structural, networked and relational nature of the global knowledge economy means that the opportunities for high-technology upscaling are increasingly reliant on a more open, quasi-market economic development strategy. As it is widely known, the ability of MNCs to engage in locational flexibility bestows upon them a certain degree of asymmetrical bargaining power in that they can choose to de-invest in a location and move their resources and technologies to alternative locations. Moreover, the increased trade integration under globalisation that has taken place in recent times means that firms seeking advanced processing services could now choose to include the rapidly reforming economies of Vietnam or Cambodia as production locations as easily as they could choose to stay in China.

This may help to explain the national leadership's concern about the future sustainability of China's current growth patterns (Irwin Crookes, 2008).

Certainly, the Chinese administration seems quite aware of the need for a revised, more collaborative-oriented, upgrading strategy. After all, it is clearly obvious that the previous policy programs such as FDI and neo-techno-nationalism had failed, and the networked nature of the global knowledge economy demands a policy shift towards a more collaborative-oriented development plan. Indeed, the most recent Chinese policy directives do suggest a move away from "forced technology" transfer policies towards the development of a new, more sophisticated, technological upscaling strategy that does not resort to techno-nationalism and instead focuses on the construction of policies that will serve to embed foreign firms in the Chinese economy via long-term R&D collaborations and alliances between Chinese and foreign-owned firms.

The key to the successful implementation of this policy plan is the ability of the Chinese government to strategically facilitate and manage the growth of sets of collaborative alliances that function to produce knowledge that is jointly owned and distributed by domestic and foreign firms. It also requires the Chinese government to think beyond the idea of domestically appropriated value.

Alliance capitalism and the Chinese global wireless communication sector

In order to participate in the development of the 4G ecosystem and gain core technology and market share, the Chinese government has undergone a shift in its political agenda, technological focus and R&D expenditure allocations. In response to socio-technological changes in the global wireless communication sector, specifically the ways in which these technologies co-evolve and require interoperability, it has also sought to forge and reinforce far-reaching global alliances with foreign and domestic multinationals. By leveraging these new structures of state-led global alliance capitalism, Chinese state leaders have been able to facilitate the favourable insertion of China's domestic firms and hence, its national economy into global production and innovation networks (Haberly, 2011). In these global innovation networks local assets are increasingly integrated into extra-territorial production systems, which allow foreign actors to assume part of the value creation of, and control over, a particular asset (Phelps, 2008; Phelps, 2007 cited in Haberly, 2011).

Furthermore, in order to ensure global system embeddedness, technological convergence and global market share, the Chinese state-led developmental policy in the sector has shifted from one of primarily promoting and protecting indigenous standards from foreign competition via regulatory mechanisms and the provision of R&D funding via a techno-nationalistic policy framework, to one that is also focused on global industry development and cooperation. This has involved the state, its key agencies and China Mobile undertaking active coordination and negotiation with major stakeholders in the 4G TD-LTE sector in order to expand TD-LTE development globally, to promote collaborative breakthroughs in key technologies and expand the membership of the industry supply chain. Indeed, the Chinese government has played a fundamental role in both the global construction and coordination of the TD-LTE full value chain itself.

The government of China has sought to play a fundamental role in shaping the development of emerging technology platforms and collaborative socio-technological institutional models in the 4G-LTE sector by acting as a key system integrator and network organiser at a global level. For example, China Mobile, in a bid to both increase its innovation footprint in the global 4G LTE global ecosystem and commercialisation processes, is engaged in collaborative behaviour at multiple levels, including inter-firm collaboration, strategic alliances, joint R&D development and the coordination of the emerging 4G LTE network via its participation and management of global initiatives such as the 3GPP. It has endeavoured to do this by constructing a global collaborative platform for the co-development of time-division duplex (TDD) and frequency-division duplex (FDD) standards and network devices in order to facilitate co-adaptive product development and value appropriation, and to ensure network embeddedness in the emerging 4G market. This involved the very quick development of anticipatory standards and a framework for the organisation of technological coordination and cooperation around the standard at a global level. The role of the Chinese state in facilitating and coordinating the development of the 4G TD-LTE ecosystem will be further examined in the next chapter.

As it was established in preceding chapters no single organisation or territory alone possesses the ability to gather whole sets of complementary knowledge, competences and resources that are needed for achieving innovative technological innovation. This situation results in a strong inter-dependency between R&D actors and innovation processes. It raises, in turn, the need for designing and implementing varied cooperative tools (alliances, partnerships, networks) and contractual devices (co-funding, joint ventures, consortia, private-public programs, etc.) that combine with more classical competitive instruments (costs and prices, product quality, marketing, etc.) in shaping the new dynamics of market competition and commercialisation and the broad diffusion of new developments. In order to participate in the contemporary global economy and its emergent production structures, supply chains, innovation chains and intellectual property regimes, it has become increasingly necessary for the Chinese government to undergo a profound shift in political economic focus. This is because in the contemporary global economy, comparative and competitive advantage is not a product of the creation of firms per se or national champions, but instead by the creation of globally linked innovative learning environments and sector-specific development.

Indeed, it is interesting to note that there are actually more foreign firms involved in the development of 4g TD-LTE than domestic ones, and these foreign firms are directly linked to China Mobile which is the focal firm and network organiser. For example, there are nine Chinese firms participating in the development of 4G TDE and 12 foreign firms. The Chinese firms participating include: Agilent Technologies Inc., Datang, Datang Microelectronics Technology, HiSilicon Technologies, Huawei, Innofidei, Leadcore Technology, Potevio and ZTE. The 12 foreign firms include: Alcatel-Lucent, Apple, Asus, Ericsson, Google, HTC, Motorola, NSN, Samsung, Sony Ericsson, SpreadCom Telecommunications and ST-Ericsson (Kwak, Lee and Chung, 2012). This suggests that the Chinese government and the state-owned China Mobile have acquired the relational and organisational legitimacy that is necessary for alliance development and ecosystem coordination at both domestic and global levels.

Indeed, intimately tied to the emergence of state-led alliance capitalism in the Chinese 4G sector is the emergence of new sets of capabilities and institutional mechanisms designed to ensure the successful construction of a global market ecosystem and the development and coordination of appropriation tools to capture key 4G value chain rewards and benefits. These include: innovation chain value appropriation such as tax collection, high-tech job facilitation and R&D cluster building. Furthermore, a number of policy shifts are occurring in China's overall regulatory framework which indicates that the Chinese government is moving towards the similar treatment of foreign and local firms. That is, the central government is seeking to gradually eliminate discriminations against foreign companies in the area of operational rights as well as removing preferential taxation offered to these companies vis-à-vis local counterparts (e.g., by combining foreign and domestic business laws into a single piece of legislation) (Moody, 2011). Moreover, recently, MOST lifted the limits on foreign investment in research areas to encourage more overseas companies to build technical centres in China. Pilot programs for private capital into the telecommunication sector may be launched in September by the MIIT (July 26). Specifically, the plan targets emerging technologies as the focus of its partnering opportunities for foreign enterprises, allocating 18 billion yuan, in R&D for eight emerging technology sectors (Moody, 2011).

Former PRC Premier Wen Jiabao recently articulated:

> As active participants in China's economic development, reform and opening-up, multinational companies want to participate in the development of those industries... [and] have said on many occasions that foreign-invested enterprises established in China in accordance with the law are Chinese companies. Their products are made-in-China products and their creation is Chinese creation. Foreign-invested enterprises in China should enjoy a fair, transparent and equal competition environment and we give them national treatment.... To develop strategic emerging industries is an important measure we have taken to meet both China's immediate and long-term needs. In developing those industries, we must be open to the outside world. We need to attract the participation of foreign companies, in particular their technology and talent. I wish to state in a serious manner that foreign-invested enterprises, when participating in the development of strategic emerging industries in China, will enjoy the same policy treatment as their Chinese counterparts. (Jiabao, 2012)

For instance, a US executive with extensive operations in China recently stated that an official from the Ministry of Foreign Trade and Cooperation told him that the indigenous innovation policies don't apply to his company because the government regards it as a Chinese company. The executive said his company felt no more discrimination selling products in China than in other nations, such as India, and that it has a fair opportunity to provide input on formation of standards. "'Indigenous innovation has been bashed down and killed for now', the executive said. 'This is something we've taken off our list as something we have to focus on'" (Wessner and Wolf, 2012).

Furthermore, another US executive said that a high-level official of MOST met with multinational representatives in June 2011 and explained that "indigenous innovation is really about improving China's ability to generate new ideas rather than displacing foreigners, and that China's innovation system is open to multinationals. The MOST official also for the first time discussed ways in which foreign companies could participate in national government-funded research projects, an opportunity many multinationals have long sought" (Wessner and Wolf, 2012).

For example, there are five core incentives offered by programs designed to spur on research and development. These incentives can be applied for by foreign-invested Chinese resident companies. They include:

(1) A reduced Corporate Income Tax (CIT) rate of 15% for new and high-technology enterprises. In order to be eligible companies most own intellectual property for the key technologies of the products and fall under one of several key sectors including electronic information technology, new energy and energy conservation, high-technology services, and others. They must also engage in R&D and incur R&D expenses, and generate profits from new and high-technology products.

(2) A reduction in the CIT rate from 25% to 15% for three consecutive years and a super deduction – 150% deduction for eligible R&D expenditures. In order to be eligible they should design new products, formulate new technical procedures, develop new skills, and have R&D expenses including but not limited to design fees, development and manufacturing costs of moulds and technical instruments, and appraisal and certification expenses.

(3) Tax concessions for advanced technology service enterprises. In order to be eligible, the company should conduct work in at least one of the following advanced technology services: information technology outsourcing (ITO), business process outsourcing (BPO) or knowledge process outsourcing (KPO).

(4) Custom duty and VAT exemptions/refunds for R&D equipment purchases. This incentive applies to certain R&D equipment purchases by qualified foreign-invested and domestic R&D centres and institutes.

(5) Tax concessions on technology transfer. (JLJ Group, 2013)

In order to gain access to the incentive program a number of leading multinational companies have adopted an "innovate with China" approach, which consists of launching R&D centres in China that focus on developing technologies for the Chinese market. General Electric, for example, established a Chinese R&D centre that focuses on developing products in line with local market demand and stated government priorities, such as rural health care and sustainable development. Siemens has a similar centre working on LED lighting products and low-cost medical equipment. Each product from these centres is tailored to the Chinese market and could potentially be sold in other developing markets. This approach serves to limit the exposure of intellectual-property risks to technologies or products developed in China. Empirical research indicates that foreign firms are now beginning to invest in China not just to access its cheap labour, but also to assess its increasingly abundant, highly skilled human capital. The OECD indicates that China is the second-largest country in terms of the number of researchers produced. This has attracted a number of multinational firms focused on utilising this highly skilled human capital. For example, China has increased R&D investment over three consecutive years for Ericsson. Mobile technology is a key R&D strength of the Chinese. Alcatel-Lucent does testing in its 3G Reality Centre in Shanghai, and NTT DoCoMo conducts 4G research in its Beijing lab. MNEs now need to adapt their products and services to local markets more than ever before, which often requires extensive local knowledge (Kozhikode, 2009). A representative from Alcatel-Lucent asserted: "To navigate China's highly competitive market, the firm has to stay agile and innovate. For example, with China Mobile, we do what we call 'co-creation.' We asked them to participate with us at a very early stage of the development of a product.... It is a different way of working together: more intense, but I think it pays off" (Alcatel-Lucent, 2012).

Chinese government complicity in intellectual property theft and industrial espionage

In order for foreign firms to engage in R&D at both basic and collaborative levels with Chinese domestic firms, China's recent history of IP theft and state-sponsored cyber espionage targeting intellectual assets and proprietary information will need to be urgently addressed. As discussed in Chapter 3, China is the biggest conductor of IP theft in the world, accounting for roughly 70% of international IP theft, according to the IP Commission Report (Wortzel, 2013). The report asserts:

> the Chinese encourage IP theft and that both business and government entities engage in this practice. According to the U.S. National Counterintelligence Executive, Chinese actors are the world's most active and persistent perpetrators of economic espionage obtaining trade secrets and continuing infringement of trademarks, copyrights,

and patents. IP are stolen from American universities, national laboratories, private think tanks, and start-up companies, as well as from the major R&D centers of multinational companies. (Wortzel, 2013)

The report cites a number of recent examples including that of American Superconductor Corporation, which had its wind-energy software code stolen by a major customer in China resulting in not only losing that customer but also 90% of its stock value. Another case cited is a US metallurgical company that lost technology to Chinese hackers which cost $1 billion and 20 years' time in development. It has also been conjectured that there exists a direct correlation between the US industries that were compromised and those industries designated by China as strategic in their 12th Five-Year Plan (2011–15). This, it is argued, supports the assertion that such attacks are indeed state-sponsored with the intention to steal IP. Chinese cyberwar poses a significant threat to the United States in terms of protecting IP, protecting critical infrastructure and maintaining economic and technological superiority (Wortzel, 2013).

However, it can be argued that as Chinese firms form more collaborative technological alliances with foreign firms and develop their own sophisticated IP in global innovation networks, the Chinese government and its intellectual property regime and rights enforcement mechanisms will have a strong incentive to develop a more trans-territorial property protection regime that is designed to ensure Chinese participation is not compromised due to the presence of state complicity in cyber-hacking cases.

Conclusion

The need for states to engage in trans-territorial technological governance in order to facilitate the co-production of technological assets and shape the emergence and trajectory of technological ecosystems has significant implications for the developmental state and its associated developmental capacities. Whereas before the state was engaged in strategic developmental projects designed to enhance the innovation and technological development of its own national firms, in the new collaborative economy, the existence of critical ecosystem dependencies and more relational nature of contemporary socio-technological ecosystems mean that it may also need to develop sets of capabilities that can also help to fund and facilitate the development of foreign ecosystem actors that are crucial to the effective functioning of the technological ecosystem. This may include fiscal, technological and regulatory support or international diplomacy such as standards and licensing negotiations.

This emergence of new forms of national state-global governance in emerging technological ecosystems has been defined as "multi-scalar" or nodal in terms of spatiality and policy reach. These formulations, too, are essentially ecological concepts, thus the concept of ecology can be extended to apply also to governance; that is, the extended, multi-actor, multi-node, "modular" network governance that is emerging globally (Salter and Faulker, 2011). As Salter (2011) asserts:

> a key element in such strategies is the need for an awareness of the emergence of new "sectors" and their steerage by nation states in interaction with transnational entities in the context of the global innovation landscape. In this context, a constructionist social theory perspective supports the insight that regulatory governance is one key driver that contributes to the defining of the boundaries of scientific and technological jurisdictions which can be supported, funded, structured, organised, standardised, contested and governed by the state. ... Novel, hybrid and combinatorial technologies present policy with the need to alter the boundaries between existing institutional arrangements and devise new administrative units.

In the case of the global wireless communication sector, the spatial and sectoral policy reach is framed around modular innovation and knowledge zones, and the existence of sets of technologies that require interoperability and technological convergence to operate. From this theoretical perspective, it is a state's ability to respond to global opportunities and the frequently transnational nature of technological innovation, rather than the coherence of its inward-looking policies, that provides the key to enhancing its comparative position in the global knowledge economy and its associated technological ecosystems. That is, "successful geopolitical maneuvering by states is more likely to be characterised by the constructive use of permeable borders than the rigid application of sovereign jurisdiction" (Salter, 2011).

Alliance capitalism, from this perspective, signals the diminishing importance of territorial economic policies in the 4G sector as the Chinese government endeavours to cultivate both public and private allies in a bid to extend Chinese economic sovereignty globally and institutionally. That is, new forms of state-led global alliance capitalism take the form of complex combinations of multiple actors and firms linked to one another through dense overlapping networks of economic and political connections. Any economic activity occurring in these networks can be said to be embedded simultaneously in multiple states, through both territorial and non-territorial channels (Haberly, 2011).

In the following chapter, I employ empirical-based data to highlight how this new, alliance-based, high-technology, developmental strategy is operating to successfully improve the positioning of Chinese domestic firms in global innovation and production networks in the wireless communication sector. I will also highlight how the Chinese developmental state as a strategic actor has developed new sets of trans-territorial policy instruments and modes of technological and economic diplomacy in order to facilitate and shape the ecosystem-building alliances in the 4G TD sector. I will argue that the success of this new globalised, alliance-based program provides strong evidence for the existence of strategic agency in the wireless technology sector. Specifically, it points to the way in which strategic agency can be achieved via collaborative ecosystem alliances, R&D expenditure, incentive programs and regulatory power.

6
Global Wireless Communication Sector

This chapter will be empirical in nature. It will focus on the global wireless communication sector in order to develop an empirical account that can highlight the emergence of "state-led" alliance capitalism in China. It will focus on China's global wireless communication sector, specifically the positioning of Chinese domestic firms in global innovation and production networks and the occurrence, nature and number of strategic R&D and ecosystem-building alliances that exist between Chinese and foreign firms in the sector. It will present empirical data relating to the increasing innovative capacity of key Chinese firms operating in the sector and highlight how these firms have managed to become firmly embedded in the contemporary and future global wireless communication sector's ecosystem through the development of innovative core technologies, the co-shaping of intellectual property standards and a focus on collaborative global innovation network development. Indeed, as this chapter will highlight, the shift from 3G to 4G technologies provides strong empirical evidence that Chinese firms that have adopted a collaborative, innovative-based strategic development policy, which views innovation as a global network approach rather than a techno-nationalistic one, are becoming network leaders with increasing intellectual property portfolios and global market share in the sector.

Methodology

Because the global wireless communication sector itself is comprised of a long-term, complex industrial chain, it was necessary to simplify the role of key actors in the sector for analytical purposes. Conversely, I have adopted the industry categorisation set out by Hallikas et al. (2005), which condenses the roles of the main network actors into five key categories, which are: equipment providers (EPs), application providers (APs), service providers (SPs), content providers (CPs) and operators. The supply-chain members outside the information communication technology (ICT) industry include the regulator, end user and intermediary. Within the wireless communication sector, EPs provide equipment and network infrastructure, handsets, etc.; APs provide applications such as software, etc.; SPs provide service platforms; CPs provide content music, pictures, mobile games, etc.; and operators are a body that the government has authorised to own the basic infrastructure through which various services can be provided to end users. Outside the ICT industry there exists the regulator, which is the organisation that devises the various industrial regulations; the end user, which is comprised of the customers who consume the devices and various services; and the intermediary, which sells handsets or SIM cards to the customer (Hallikas, 2005 cited in Lei et al., 2008).

The empirical focus is here is primarily confined to the strategic, innovation-based behaviour and alliance data of EPs, SPs, operators and regulators.

EPs and APs are responsible for the development of new technologies within the ICT supply chain. Therefore, the mastery of the core technologies or standardisations is the most important value-adding opportunity for EPs and APs (Lei et al., 2008). Huawei and ZTE both operate in this category.

Network operators occupy a key position in the ecosystem precisely because they possess critical resources: mobile networks (Lei et al., 2008). There are currently only three mobile network operators in China: China Mobile, China Unicom and China Telecom. Estimates suggest that mobile operators in China are 70% state-owned (Zhang and Prybutok, 2005). This book will only examine the behaviour and policy directives of China Mobile. This is because China Mobile is the only Chinese operator allocated 4G TD-LTED spectrum licenses. As a direct result, it is the role of China Mobile to facilitate network and ecosystem development in the sector around TD-LTED technologies.

It is important to note that because the Chinese government is both the regulator and the owner of the three major telecommunication mobile network operators in China, it exercises central control over China telecommunication policy making. That is, China's telecommunication sector at the operator level is primarily state-owned and directed. In China, MIIT is the official regulator of telecommunication and all information-technology-related markets in China. MIIT reports to the State Council. Although the MIIT has relinquished direct control

of all owned telecommunication operators and equipment manufacturers, it retains significant influence over the whole sector, as most
executives working in the sector have links with MIIT. MIIT's functions
include: developing a national telecommunication policy framework
implementing sector regulation for manufacturing, production, service
and software; planning public networks, including local, backbone and
Internet; regulating private networks; developing technical specifications and performance criteria for products used in the public networks;
allocating frequencies for use in wireless services; approving a new
frequency spectrum and services; and developing postal and telecommunication service-charging policies. MIIT also represents China in
signing relevant regional and international treaties for Chinese/foreign
cooperation. MIIT faces inherent conflicts of interest, as it seeks to foster
the development of China's telecom equipment industry and to oversee
Chinese operators that are state-owned (Lei et al., 2008).

The Chinese government: learning from the failure of TD-SCDMA

It has already been established in earlier chapters that the early development of policy directives in the Chinese information-communication
sector involved concerted attempts by the Chinese government to
curtail the dominance of foreign firms. It also sought to foster independent innovation in a bid to facilitate the movement of Chinese
domestic firms up the sector's value chain. Specifically, this involved
the issuing of distinctive standards for a range of 3G cellular telephony
technologies in the ICT sector in China, specifically TD-SCDMA and
WAPI (Ran, 2011).

Furthermore, as it has already been established, both of these standards meet with concerted foreign resistance. However, in an interesting
and theoretically pertinent twist of fate, both have become international standards – this time with the support of US and European
companies who previously opposed them. The reasons for this switch
are complex and numerous. Obviously, government procurement and
the adoption of the standards by Chinese carriers means that it is in the
interest of foreign equipment makers to produce products that contain
the standards in order to increase sales in the Chinese domestic market.
However, TD-SCDMA also failed to commercialise due to the fact that
it was an immature technology providing inferior technologies in
comparison to WCDMA, and its use was limited to within China. As
a consequence, despite its recognition as an international standard,

TD-SCDMA was not universally embraced, due to a lack of available devices, early technical challenges and its failure to move away from the shadows of WCDMA (Ran, 2011; Zhang and Liang, 2011; Ge, 2011; Feng, 2011; Yan, 2007).

Throughout this time frame, incidences of regulatory control over network operators and foreign equipment manufacturers are clearly evident. For example, with the development of TD-SCDMA, China Mobile, as a network operator, acted as a mediator between the network of firms that operated in the sector (Ligang, 2011). It sought to facilitate the building of the ecosystem's architecture and ensure that the commercialisation of TD-SCDMA resulted in the development of innovative services and products. Confronted with a lack of TD-SCDMA handsets in the market in this period, China Mobile endeavoured to force its foreign partners, Nokia and Motorola, to produce TD-SCDMA products by threatening to withdraw market access. Thus, Nokia, HTC, Samsung and other local firms produced a number of TD-SCDMA handsets (Tsai and Wang, 2011). That is, the development of devices could not occur without political lobbying and economic coercion by the Chinese central government.

Indeed, the only Chinese standards that have been successful have been those where the Chinese promoters have built broad alliance-based coalitions composed of both Chinese and foreign industry partners who could cooperate on product development and commercialisation. Thus, as outlined in earlier chapters, it can be argued that one of the key lessons of WAPI and TD-SCDMA is that in the contemporary networked economy it is not possible for Chinese policymakers or Chinese domestic firms to insert themselves within or create a global wireless ecosystem without domestic and international collaboration.

The fact that these technologies failed to successfully commercialise, coupled with an increasing awareness of the importance of global ecosystem development and the need for collaborative R&D and intellectual property sharing, has led to a shift in policy orientation by Chinese policymakers from a techno-nationalistic position to one more characterised by techno-globalism. As Datang Telecom's Vice President Chen Shanzhi recently stated: "TD has entered the pull of market period. ... In the beginning of Twelfth Five-Year, we will prepare well for the commercialisation of TD as well as the industrialisation of 4G, and go globally in TD commercial processes. In the past we say technology patented, patent standardisation, standard industrialisation, industry marketisation, today we must try our best to promote the market internationalisation" (Chen, 2011).

As this chapter will highlight, the overarching result of this emergent policy position and the subsequent change in Chinese domestic firm behaviour has resulted in the development of standards that both anticipate and operate to define the future regulatory global infrastructure and market governance structure. It has involved the development of network and research alliances between Chinese domestic firms and foreign stakeholders and actors. As a direct result, in this the round of ecosystem development the Chinese state as a wireless technology ecosystem actor and key Chinese domestic firms are well positioned to play a fundamental role in shaping the future regulatory and market dynamics in the 4G global wireless communication sector.

4G

Short for fourth-generation wireless communication systems, 4G has engaged the attention of wireless operators, equipment makers, original equipment manufacturers (OEMs), investors and industry watchers around the world. 4G refers to the next generation of wireless technology that promises higher data rates and expanded multimedia services. "The transition from 3G to 4G will lead to a dramatic increase in data transmission speed from two megabits per second (mpbs) to 1,000 mbps. If 3G took mobile phones to the threshold of images and video, 4G will bring them into an era of the superhighway where downloading full-length films will require only seconds" (Caijing, 2009).

At present, China has defined TD-LTE as one of the top priorities of the "11th Five-Year Plan" in its key program of "Next-generation of Broadband Mobile Communication Network" (Ge, 2011; Guo, 2006; Ligang, 2011; Zang, 2011; Feng, 2011; C114, 2010). China Mobile owns 2.3 GHz and 2.6 GHz TDD bands and is a leading promoter of TD-LTE (Min, 2011). Indeed, it has been conjectured that China will be the main driver for the deployment of the 4G mobile telephone network in the Asia-Pacific region and that there will be will be 126 million subscribers by 2015 with China being home to 58 million of them (Rajaraman, 2011). As Jaikishan Rajaraman, senior director of GSM Association's Asia-Pacific branch argues: "We believe 2011 will be a tipping point for the LTE industry, as 24 countries are about to build up LTE networks by the end of this year...there will be 4 million LTE subscribers globally by then....The 4G network will really take off in 2012, and by 2015, the number of users is expected to rise to 300 million across 55 countries" (Rajaraman, 2011).

China Mobile Communications Corporation, parent company of the Hong Kong-listed China Mobile Limited, has engaged in pre-competitive market construction since 2007 to prepare for the rollout of 4G commercial services in China, by adopting TD-LTE technology. TD-LTE, also referred to as LTE-TDD, is a mobile telecommunications technology and standard that has been co-developed by Datang Telecom, China Mobile, Huawei, ZTE, Nokia Siemens Networks, Alcatel-Lucent Shanghai Bell, Qualcomm, ST-Ericsson, Leadcore and others (Motorola, 2010). It is one of two variants of the 3GPP LTE technology, the other being FD-LTE (or LTE-FDD). Likewise, TD-LTE-Advanced is an LTE advanced time-division variant, an evolutionary upgrade version of TD-LTE. TD-LTE-Advanced can easily reach download speeds of more than 150 megabytes per second, much faster than 3G TD-SCDMA technology. This upgraded version of TD-LTE (TD-LTE-Advanced) is now among the three international 4G standards accepted by the United Nations (UN) International Telecommunication Union (ITU). The other two are LTE-FDD and Worldwide Interoperability for Microwave Access (WiMAX), which are dominated by Europe and the United States, respectively (Motorola, 2010). It has been estimated that TD-LTE and LTE-FDD will coexist with global shares of 33% and 67%, respectively (Goldman Sachs, 2011).

It is important to note here that TD-LTE is often wrongly described as a Chinese technology. In fact TD-LTE shares approximately 90% of its core technology with "mainstream" LTE-FDD (Grivolas, 2011). The differences between TDD and FDD are solely a physical layer manifestation and therefore invisible to higher layers (Motorola, 2010). As a result, there are no operational differences between the two modes in the system architecture. Indeed, the key advantage of the LTE ecosystem is the vast economy of scale gained through combining LTE-FDD and -TDD (TD-LTE) in a standardised way. The 3GPP standard allows both devices and implementations to be simpler, a major factor in reducing cost for deploying a mobile broadband technology. Since other TDD-based technologies, including TD-SDMA and WiMAX, also have a migration path to LTE, combining FDD and TDD makes the scale and economy of the LTE ecosystem very attractive to operators (Grivolas, 2011).

Thus, with support from the Chinese central government, China Mobile developed a plan to capitalise on LTE network congestion, which is becoming an increasing problem as data traffic levels rise rapidly. Operating from the premise that TDD as an unpaired spectrum will trade at a significantly lower price per MHz, as well as the fact that many countries throughout the world have large chunks of unpaired TDD spectrum available – both used and unused – China Mobile, in a clear example of institutional entrepreneurship and anticipatory governance, has sought technological leadership in the sector as a powerful network coordinator and ecosystem builder.

It has endeavoured to do this by constructing a global collaborative platform for the co-development of TDD and FDD standards and network devices in order to appropriate value and network embeddedness in the emerging 4G market. This involved very quick development of anticipatory standards and a framework for the organisation of technological coordination and cooperation around standards at a global level. That is, China Mobile has strategically sought to position LTE-TDD as the de facto global standard for mobile broadband – something most cellular operators would welcome for cost reasons (Grivolas, 2011:1). As China Mobile's (former) Chairman Wang Jianzhou asserted: "The development of mobile broadband highlights the unique flexible advantage of TDD spectrum... China Mobile has been committed to the converged development of LTE-TDD and FDD, which will achieve the economy of scale and the wide support from global industry. With great effort from all partners, TD-LTE enters commercialisation and global deployment... Clearly this is not just a Chinese technology" (Wang, 2011).

Moreover, press releases, conference proceedings and policy documents further support the assertion that the Chinese government has adjusted its previous indigenous technology policy framework in order to adapt to the collaborative nature of contemporary technological ecosystems. For instance, Stephen Hire, director of marketing at testing and measurement company, Aeroflex Asia, speaking at the next-generation mobile broadband session at CommunicAsia, asserts that the Chinese government has applied lessons learned from its 3G standard (TD-SCDMA) on how to proceed on the development and promotion of TD-LTE and redesigned its policy directives to incorporate the interests and ecosystem dependencies of foreign firms. He further states:

> One of the weaknesses of TD-SCDMA was that many saw it as too far separated from the core 3G technology, so there was very little international participation......in contrast, TD-LTE is part of the global core LTE community.... With limited participation, economies of scale around TD-SCDMA could not be achieved.... On the other hand, it was easier for manufacturers to develop for TD-LTE over the common chipset, and this reduced technical hurdles and lowered the barriers of entry for manufacturers.... The Chinese government's strong commitment to promote TD-LTE also played a part in providing incentives for foreign companies to invest in the technology.... For instance... it channeled funds toward supporting

government-sponsored organisations, such as the TD Industrial Alliance, to develop the TD-LTE ecosystem. (Hire cited in Qing, 2011:20)

In a similar vein, China Mobile's General Manager of Technology Zhou Jianming, speaking at the 4G World China Conference in Beijing, asserts: "Indeed, the development and use of TD-LTE Advanced will be shaped largely by the market. Within China both TD-LTE and FD-LTE will be supported by Chinese government and international firms....China Mobile hopes consumers will be able to access both the TD-LTE network and the LTE-FDD network with the same handset....Convergence will provide a much bigger platform for TD-LTE technology" (Zhou, 2011; Feng, 2011).

This example serves to further support the argument that Chinese high-technology policymakers have moved beyond the idea of techno-nationalism and the assumption that the only way to achieve significant market share in the global wireless communication sector is via first-mover status and indigenous standard ownership. Instead, China Mobile has sought to shape and construct a new technological ecosystem via collaborative alliances and network coordination. The overarching goal here was to enlarge the market itself by offering a niche product that exploits network congestion. It highlights how the Chinese government via the state-owned China Mobile is engaged in innovation ecosystem-building alliances and the recruitment and development of other firms and system actors as a strategic way to ensure that critical infrastructure, regulatory standards and complementary technologies are developed and critical ecosystem dependencies are addressed, in order to secure the successful globalised commercialisation of TD-LTE wireless technology.

China's domestic firms and evidence of innovation capability

Empirical data and interview transcripts indicate that two key Chinese domestic firms, Huawei and ZTE, and the state-owned China Mobile have become firmly embedded in the LTE-Advanced and TD-LTE global wireless ecosystem. All three firms are becoming network leaders, with a range of dynamic capabilities and collaborative network linkages and alliances.

Acquiring innovation capability is used here to refer to improving the ability for innovation and self-developed technologies, which is in direct contrast to the strategies of imitating or assimilating obsolete technologies of more advanced countries. Also important is the capacity to produce knowledge in networks and create value from networks. A true innovation network is defined by partners who collaboratively develop innovative intellectual property, then proliferate the innovation throughout the network and to the network members, partners and customer ecosystems (Zutshi, 2009). This data will be presented in four subsections. These are alliance data, R&D expenditure, patent data, and market share.

Alliance data and R&D expenditure

Evidence that Chinese domestic firms and the Chinese government are increasingly engaging in collaborative R&D, ecosystem-building alliances and technological system shaping in order to become firmly embedded in global production and innovation networks can be drawn from recent data on R&D partnering. Three key types of alliances can be isolated for analysis: R&D alliances, standard alliances and market alliances. All examined Chinese firms engaged in a mix of all three alliance types, with varying forms and levels of inter-firm commitment.

China Mobile

China Mobile is the world's largest telecom carrier by subscriber numbers. As of November 2011, China Mobile had 638.89 million subscribers (Lo and Lewis, 2011). In 2011, China Mobile was cited as the number one telecommunication firm in the world by revenue. The company ranked 10th in the *Financial Times'* Global 500. China Mobile's position on the *Forbes* Global 2000 list rose from 55th to 38th place in 2011. The China Mobile brand was named one of "BrandZ™ Top 100 Most Powerful Brands" by Millward Brown and the *Financial Times* for the fifth consecutive year, and the brand value currently ranks 8th, topping all other telecommunications operators in the world (China Mobile, 2010).

As it has already been established, China Mobile as a network organiser has been responsible for stimulating innovation and overseeing the coordination of the Chinese wireless communication 4G sector's ecosystem. In order to do this it is required to establish relationships with SPs, CPs and device manufacturers, as well as equipment and systems providers. China Mobile has asserted that its primary developmental strategy for the group is a focus on open-platform operation and the adherence to cooperative and win-win principles (China Mobile, 2010).

A key feature of China Mobile's developing business model for the TD-LTE sector is the shift from an overt focus on the development of indigenous knowledge and standards development to one of global market maker and coordinator. This has resulted in an expanded role for China Mobile as an alliance facilitator and coordinator at the global level. One example of China Mobile's changing business model and the Chinese government's evolving high-technology development policy can be drawn from China Mobile's role in setting up the Global TD-LTE Initiative (GTI) which was launched in February 2011 by China Mobile, Bharti Airtel, Softbank Mobile, Vodafone, Clearwire, E-Plus and Aero2. The primary goal of the organisation, which as of December 2011 possessed 35 members, is to focus on facilitating multilateral coopera-tion among members in an attempt to promote the fast development of TD-LTE technology, the convergence of TD-LTE and FDD modes to maximise economies of scale, and the sharing of the ecosystem with other TDD technologies, such as the Japanese extended global platform (XGP) technology. That is, GTI is designed to create value for stake-holders across the TD-LTE ecosystem, facilitate the early adoption of the technology, and ensure the convergence and interoperability of TD-LTE and LTE-FDD. As the general manager of China Mobile Research Institute, Huang Xiaoqing asserts: "Operators worldwide are looking for a single standardised protocol" (Huang, 2011b).

Furthermore, China Mobile is also engaged in a network of R&D alli-ances with Sony Ericsson, Nokia Siemens, Huawei, ZTE, Datang, Alcatel-Lucent, SK Telecom, Apple, Google, Rohde & Schwarz, and Clearwire in a bid to facilitate the production of network devices. To promote TD-LTE devices, China Mobile's (former) Chairman Wang Jianzhou stated that China Mobile will enhance cooperation with Taiwanese hardware partners, including HTC Corp, Foxconn Technology Group and MediaTek (GTI, 2011). In 2011, two top Chinese and South Korean operators – China Mobile and SK Telecom (SKT) – forged closer links with a deal that will see them jointly develop technologies focusing on next-generation networks, device platforms, machine-to-machine serv-ices and the booming applications market. Moreover, China Mobile has established an open interoperability testing lab with seven system vendors and six terminal chip vendor participants to date (Ayvazian, 2011:10).

China Mobile has conducted extensive R&D lab and field trials and has been working closely with operator partners such as Vodaphone and

Verizon and international organisations such as 3GPP, Next Generation Mobile Networks (NGMN), LTE/SAI Trail initiative (LSTI), the Global System for Mobile Communications (GSM) and other key organisations and actors to promote LTE-FDD/TDD convergence and synchronised development. Moreover, China Mobile is currently cooperating with test instrument manufacturers, such as Rohde & Schwarz, Anite, Agilent and Anritsu, in order to develop test cases. Cao Shumin, (now) President of the Telecommunication Research Institute (TRI) of MIIT, has recently emphasised the global nature of the TD-LTE standard and the open and inclusive nature of the testing process, asserting: "tests on the network are fully open...the test site at the MIIT institution has gathered not only domestic cell phone manufacturers but also multinational tycoons like Motorola, Ericsson, and Nokia Siemens Networks....But as the two technologies are based on the same LTE system, they are able to share R&D results and subscribers at a global level" (Cao, 2011). China Mobile is also a member of key standards bodies and took a leading role in 36 industry-standards development projects, participated in 63 projects and completed the establishment of 26 industry standards in 2010 (China Mobile, 2010).

Indeed, China Mobile provides us with an interesting case of state-led alliance capitalism in China. For instance, as a state-governed operator, China Mobile highlights how a state can be successful at technological ecosystem building and market coordination and facilitation at a global level and affirms the willingness of the Chinese central government to adapt its economic policy making towards a more collaborative alliance-based model of technological development. Furthermore, it is also interesting to note that for the first time China Mobile has engaged in an attempt to approach standardisation in a bottom-up fashion from the market, rather than top-down from the government (Ernst, 2010). In this way, it has been suggested, that we may be witnessing China Mobile's first steps towards its accession as a global mobile operator. As the chairman of China Mobile asserts:

> realise that new model is required the second opportunity is the industrial chain collaboration, which is more important than ever....I talked with U.S. AT&T's CEO yesterday, and we both feel that the most difficult for us today is not the technology, money, market, but how to adapt the changes of our business model and ecological chain, which is the biggest challenge to operators...this is the first time Chinese mobile communication industry can stand on the same starting line with peers in developed countries. (Sha, 2010a)

Huawei

Founded in 1988, Huawei is a privately owned company owned directly by its employees. It is located in Shenzhen, China, and is currently the largest Chinese telecommunication equipment maker. Its core activities include building next-generation telecommunication networks. providing operational and consulting services and equipment to enterprises, and manufacturing communications devices for the consumer market (Huawei, 2011).

Huawei has engaged in collaborative R&D and networking ecosystem development via a number of partnerships and alliance forms since its inception. In 2000, Huawei partnered with IBM in order to gain access to IBM's network processes' R&D technology centres (Ahrens, 2013). In 2003, Huawei, NEC and Matsushita (Panasonic) established a joint venture (JV) company, Cosmobile, to share smartphone technology. Later that year, NEC and Huawei opened the 3G Mobile Internet Open Lab to incubate 4G technologies (Ahrens, 2013). In March 2003, Huawei and 3Com Corporation formed a JV company, 3Com-Huawei (H3C), which focused on R&D production and the sales of data-networking products. In 2005, Huawei and Siemens formed a JV called TD Tech, which focused on the development of 3G and TD-SCDMA mobile communication technology products. The US$100 million investment gave the company a 49% stake in the venture, while Siemens held a 51% stake. In 2007, after Nokia and Siemens co-founded Nokia-Siemens Networks, Siemens transferred all shares it held in TD Tech to Nokia Siemens Networks. In 2007, Huawei and American security firm Symantec announced the formation of a JV company to develop security and storage solutions to market to telecommunication carriers. Huawei owns 51% of the new company, named Huawei Symantec Inc., whilst Symantec owns the rest. In 2008, Huawei launched a JV with UK-based marine engineering company, Global Marine Systems, to deliver undersea network equipment and related services. Later that year, Huawei also established a JV with Telecom Venezuela, called Industria Electrónica Orinoquia, for R&D and the sale of telecommunication terminals. Telecom Venezuela holds a 65% stake while Huawei holds the remaining 35% stake. In October 2010, Sequans and Huawei entered into a partnership to develop and mature TD-LTED technology for the global marketplace. In 2010, Huawei set up a joint innovation centre with Telenor in Pakistan. In 2011, Huawei announced plans to set up a global research and development centre for R&D in Italy and a joint innovation centre with Bell Canada (Huawei, 2011).

Huawei actively participates in standard alliances and has part-nered with key international wireless standardisation organisations to drive technology development and the improvement of stand-ards within the industry, including 3GPP, Asia-Pacific Telecommunity (APT), Association of Radio Industries and Businesses (ARIB), European Telecommunication Standards Institute (ETSI), Institute of Electrical and Electronics Engineers (IEEE), Internet Engineering Task Force (IETF), International Telecommunication Union (ITU), Telecommunication Industry Association (TIA) and Wireless World Research Forum (WWRF). Huawei currently holds significant roles within these organisations, with 83 key positions, from chairmen and board members to rappor-teurs (Huawei, 2011).

Huawei has also sought to engage in a number of alliances to facilitate and develop market access. For example, it entered into a strategic coop-eration alliance with Marconi by reselling each other's products in their home countries (Huawei, 2011).

ZTE

ZTE Corporation (formerly Zhongxing Telecommunication Equipment Corporation) is a Chinese multinational telecommunications equip-ment and systems company headquartered in Shenzhen, China. ZTE was founded in 1985 by a group of state-owned enterprises but operates as a "private-operating" economic entity. As of 2011, ZTE is the second-largest Chinese telecom equipment maker and the world's fourth-largest mobile phone manufacturer (ZTE, 2011).

ZTE has partnered with over 230 major carriers and distributors in more than 160 countries and regions around the world and has strategic partnerships with 47 of the global top 50 carriers. In 2011, ZTE applied for more international patents than any other company in the world (ZTE, 2011). For example, ZTE has established strategic cooperation agreements with leading telecommunication giants such as Portugal Telecom, France Telecom, Alcatel, Ericsson and Nortel in NGN and mobile systems, with Hutchison in 3G, with Orchard in marketing and content bundling, and with Marconi in optical transmission systems (ZTE, 2011). In an example of collaborative platform development, ZTE announced a strategic alli-ance with Intel in March 2013, which will see ZTE equip its upcoming smartphones with Intel's newest processor, the Atom Z2580. This is not the first time that the two companies have cooperated on smartphone development. In August 2012, ZTE launched Grand X IN, the compa-ny's first Intel-powered smartphone. The Android-based device became the best-selling smartphone in Austria. It has also had successful sales in Germany, Poland, Hungary, Romania, Serbia, Macedonia, Slovakia, Moldova, Greece, Sweden and Norway, and soon will also be offered in the French market. As Ao Wen, general manager of ZTE mobile devices in Europe, articulates: "'ZTE GRAND X IN' is the first flagship smartphone powered by Intel chips in Europe...Strong cooperation with Intel, we released the product successfully in Europe and thereby improve our in brand awareness in these important markets, and we look forward to continue to expand its cooperation with Intel to support the development of the resurgence in the high-end smart phone market, Intel is the strategic partners we have developed a new state-of-the-art mobile technology" (ZDNET, 2013).

The company is an active member of more than 70 international standardisation organisations and forums. ZTE holds the position of co-chairman in two Telecommunication Standardisation Sector (ITU-T) working groups and is the editor of a number of ITU-T standards including next-generation network (NGN) and optical transmission and network security. As a member of 3GPP, ZTE has edited three standards involving 3G radio access networks and terminal systems. The company is also a board member of the WiMAX Forum (ZTE, 2011). In 2013, ZTE joined the Power Matters Alliance (PMA) to help facilitate the standardisation and convergence of wireless charging guidelines and products (Hwang, 2013). It is also a member of the Next Generation Mobile Networks (NGMN). NGMN is a mobile telecommunications association of mobile operators, vendors, manufacturers and research institutes, formed in 2006. It was founded by major mobile operators as an open forum to evaluate candidate technologies to develop a common view of solutions for the next evolution of wireless networks. Its objective is to ensure the successful commercial launch of future mobile broadband networks through a roadmap for technology and user-friendly trials and ecosystem convergence (NGMN, 2006).

ZTE is actively cooperating with various parties in the TD-LTE ecosystem. The company collaborates on interoperability tests (IOT) with mainstream chipset providers such as Qualcomm, Innofidei, Altair, Sequans and ST-Ericsson (ZTE, 2011). It also works to promote the development of the TD-LTE platform and standards. In October 2009, ZTE aligned with Innofidei to successfully demonstrate the industry's first multivendor TD-LTE high-definition video service at the ITU Telecom World 2010 conference in Geneva. At the end of 2010, ZTE was the first in the industry to launch Qualcomm chipset-based TD-LTE terminals and achieved a download rate of nearly 100 Mbps (ZTE, 2011).

R&D expenditure

A growing body of research indicates that a firm's level of R&D expenditure is a key determinant of its innovative capacity and ability to gain market share. In terms of R&D expenditures, patents and venture capital investment, the sector exceeds other industries by a significant margin (OECD, 2009). Furthermore, as well as a commitment to high levels of R&D expenditure, contemporary firms operating in the global wireless communication sector need to establish and maintain both domestic and global R&D networks that allow for research collaboration and knowledge exchange. Evidence gathered by this book indicates that China Mobile, Huawei and ZTE have all firmly established global R&D networks and high rates of global R&D embeddedness.

Table 6.1 shows a more detailed investigation of the R&D expenditure of select companies. Domestic leading firms, such as Huawei and ZTE, generally spent over 10% of revenue in R&D in 2010. This level is comparable to international R&D spending in the telecom equipment industry (i.e., from 10% to 20%). Furthermore, Huawei and ZTE both achieved an R&D staff/total employment ratio of over 40%.

All three Chinese companies have set up joint R&D labs with foreign firms in order to secure global R&D embeddedness, access to global markets and localised knowledge content. As of December 2011, Huawei Technologies has established a total of 20 joint R&D labs with: Texas Instruments, Motorola, IBM, Intel, Agere Systems, SunMicrosystems, Altera, Qualcomm, Infineon, Microsoft, Telenor, NEC and Bell Canada, among others (Huawei, 2011). "We have established more than 20 joint innovation centers with global leading operators and now lead the strong development of the LTE industry by transforming advanced technologies into customers competitiveness and success in business" (Huawei, 2011).

Table 6.1   R&D expenditure of select companies, in billions (2010)

	Expenses (US$)	As percentage of net sales
Ericsson	29.9	14.7
Alcatel-Lucent (French)	2.6	15.6
Qualcomm	2.5	23
Nokia Siemens	27	16.5
Cisco	5.2	13.6
Huawei	16.5	8.9
ZTE	7.0	10.1

Sources: Huawei, ZTE, Ericsson, Alcatel-Lucent, Qualcomm, Nokia Siemens, Cisco, company annual reports for 2010 and various public sources.

ZTE has established 15 R&D centres and institutes across North America, Europe and Asia. The company has also established joint laboratory partnerships with Texas Instruments, Intel, Agere Systems, HHNEC, IBM, Microsoft (China), Qualcomm, Hua Hong NEC, and Tsinghua University. The company has undertaken technological research alliance projects with 50 academic institutions throughout China, where ZTE is also a full-fledged member of the China Communications Standardisation Association (CCSA) (ZTE, 2011).

In 2009, China Mobile announced plans to establish an R&D facility in Silicon Valley: "This is the first overseas research and development facility that China Mobile has set up. The president of China Mobile's Institute of Research, Huang Xiaoqing, told the news site that it sees most of its revenue today coming from voice services, but the company recognises that data services are the future. And it's looking to Silicon Valley for innovation" (China Mobile, 2011b).

Foreign firms as strategic insiders

There are a number of indications that the Chinese government is not just funding the innovation programs of its domestic firms. Instead, it is also funding research collaboration between its domestic firms and foreign MNCs. Hence, the strategic policy directives of the Chinese government are now beginning to extend beyond the boundaries of its domestic firms. For instance, several interviewees noted that the Chinese government has aggressively funded efforts to promote collaboration between Chinese universities and industrial enterprises, including foreign firms. Taking advantage of this, many Western firms have been able to supplement their own research efforts by encouraging Chinese engineering and science professors to undertake R&D related to the needs of Western companies, with most of the costs being underwritten by Chinese taxpayers (Moody, 2011; Ran, 2011; Ligang, 2011; Zhang, 2011; Ge, 2011; Feng, 2011).

Some analysts believe that the multinationals working on research in China have the potential to create a fusion which can drive technological development. As Arding Hsu, senior vice-president of German electronics and engineering company Siemens, which has major R&D facilities in China, articulates: "China is an ideal place for fusion innovation to thrive. On the one hand, China's fast-growing market is characterised by massive and diverse needs.... On the other hand, the long history and brilliant culture of this country have accumulated a lot of interesting knowledge" (Moody, 2011). Chuang Ching, executive vice president of product and R&D at Alcatel-Lucent Shanghai Bell, also emphasises that government support of R&D investment is attractive for multinationals: "The government is also keen to offer incentives for foreign companies to set up R&D bases in the country as part of its 12th Five-Year Plan (2011–2015). ... The government's Five-Year Plan is focused on innovation and a very important part of that is incentives to encourage research and development investment not just for domestic but for multinational companies too, relating to costs and business tax incentives to set up research and development centers" (Chuang, cited in Moody, 2011).

Furthermore, foreign firms are increasingly deciding to conduct innovative R&D research in China with more than 41% of R&D centres focused on the ICT industry. For example, there are now 1200 foreign multinational R&D centres in China, representing a $12.8 billion investment. Indeed, of the 2400 patents awarded to telecommunications giant Alcatel-Lucent in 2010, 280 came from China, 11.75% of the global total. Chuang said China was at the forefront of the company's research and development efforts: "China has a rich talent pool and our workforce is very young, dynamic, highly motivated and able to meet challenging targets, he said. Foreign multinationals increasingly see China as the land of invention" (Chuang cited in Moody, 2011).

Patent data

One measure of strength in innovation capability, R&D output and the ownership of core intellectual property can be measured by patent filing. The filing and ownership of patents is often cited as a measure for a knowledge-based economy and a barometer to judge the spread of innovation-based companies in each country (Guangzhou LiLon Consulting & Service Co., Ltd, 2011).

A patent is an exclusive right to exploit (make, use, sell, or import) an invention over a limited period of time (20 years from filing) within the country where the application is made. Patents are granted for inventions which are novel, inventive (non-obvious) and have an industrial (useful) application (Tabarrok, 2002). Essential patents are important here; these are patents which disclose and claim one or more inventions that are required to practice a given industry standard (Updegrove, 2007). Mobile phone IPR licensing significantly includes patents that are "essential" to implement various standards including GSM, CDMA, High Speed Packet Access (HSPA), and LTE. Those are patents that are core to the 4G technology standard. LTE and TD-LTE are currently viewed as among the most valuable intellectual property in the mobile industry, due to the fact that most major wireless operators around the world are building out LTE networks (Mallinson, 2011).

As globalisation has increased technology-based competition, the key to competitive success is a broad portfolio of "essential patents" which are necessary to produce any product that meets the specifications defined in the standard. It is important to note here that the current ownership of LTE essential patents is highly contested and no consensus exists on who exactly owns key patents in the sector. Certainly, any analysis of patent data is faced with a number of methodological problems that are widely discussed in the literature. Indeed, valuing patents is not as simple as counting the number of patents a company has. Deciding which patents are "essential" is equally complicated. Misek argues that patents can be subject to special conditions that affect their worth, such as cross-licensing agreements, foreign ownership and asset-transfer restrictions. (Woyke, 2011). Indeed, it is interesting and pertinent to note that different studies conducted to establish essential patent ownership in the sector have produced varying results even when a similar methodological framework was employed. For example, research conducted by Nokia and Jeffery on patent ownership found that the proportion of patents judged essential by different researchers varied by more than a factor of ten (Mallinson, 2011). Certainly, it is not possible to independently determine the ownership of essential patents. Hence, the only credible medium for establishing the ownership and distribution of essential patents is to have primary information of the cross-licensing that companies are signing and the actual patents involved. Even then, a complete picture of patent ownership in the sector can take years to evolve, as litigation involving patent ownership is the only legal way to verify a company's claim to ownership (Mallinson, 2011).

As Figure 6.1 indicates, both Huawei and ZTE are capturing an increasing share of LTE essential patents with intellectual property databases, indicating that Huawei and ZTE owned 8% and 7%, respectively, of LTE essential patents as of November 2011. InterDigital is the leading holder of essential patents in LTE, with its patent holdings arm controlling 13% and its technology unit, 11%. Next comes Qualcomm with 13%, Nokia and Samsung with 9% each, Ericsson and Huawei with 8%, ZTE at 7%, LG with 6%, and NTT DoCoMo with 5%, while the remaining 11% is held by "others" (ETSI IPR Online Database, 2011).

Furthermore, as Table 6.2 (below) highlights, in 2011 both Huawei and ZTE are listed in the top 10 for international patent application

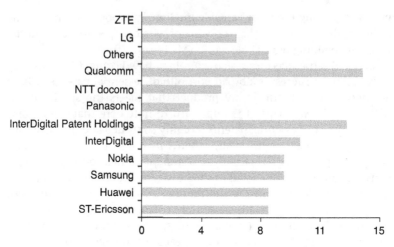

Figure 6.1   LTE essential patent ownership (2011)
Source: ETSI IPR Online Database (2011).

Table 6.2   Leading corporate patent filers (2010)

Rank	Company	Patents
1	Panasonic	2154
2	ZTE	1863
3	Qualcomm	1667
4	Huawei	1528
5	Philips	1435

Source: ETSI IPR Online Database (2011).

filers, with ZTE filing 1863 patents and Huawei filing 1528 patents, resulting in a rank of second and fourth, respectively (ETSI IPR Online Database, 2011).

Whilst the exact number of contracts and essential patents awarded to Huawei and ZTE are a matter of contention, both companies assert that their commitment to R&D has played a significant role in their increasing ownership of essential LTE patents. As a ZTE representative articulates in this statement:

The Company's share of LTE essential patents is a result of ZTE's continuous R&D investment into LTE related key technologies since 2004. ... It said it would commit more R&D resources to accumulate essential patents and adopt a proactive approach in improving 4G and developing 4G+ standards. With LTE essential patents being evenly distributed among different vendors, we believe that it will be helpful to provide a comparatively healthy and balanced IPR licensing environment for the development of the 4G industry. ... Furthermore, the firm aims to continue to file or acquire patents and increase its share of the evolving list of LTE essential IPR to 10% by 2012. (ZTE, 2011)

In a similar fashion, Huawei has emphasised how its commitment to investing in R&D to create competitive products and solutions is the reason behind its accession to a first-tier global telecommunications firm. For example, Leo Sun, chief executive of Huawei France, who is in charge of Huawei's European development strategy, asserts that: "Huawei will rely largely on its cutting edge in R&D to help achieve its goals. ... We have more than 18,000 patents and almost 50% of our employees are dedicated for R&D. In other words, we have the biggest R&D force in the world. Thanks to that innovation, we can maintain our global leading edge in innovative technology" (Sun, 2011).

Market share

Another important indicator of a firm's innovative capacity is market share. The translation of patent ownership to that of market share is a key indicator of the overall success and innovative capacity of a company. Certainly, it is well documented in the literature that without the commercialisation of technology no value can be captured, and R&D investment is lost (Ernst, 2003). Furthermore, it is important to note here that the long-term development of market share and value capturing is not only driven by price but also by technology competition and factors such as speed to market and intellectual property protection.

Furthermore, the presence of an imitation lag (i.e., the time and costs for a follower to absorb superior technology and apply it to manufacturing processes) is one important strategic pathway that can be adopted by innovative firms intent on acquiring and sustaining market share. Under these conditions, new or advanced products integrating superior technologies will form oligopolistic markets, at least temporarily, before the followers catch up (Nehausler et al., 2011). Therefore, firms developing new products integrating superior technology will dominate the markets for these products, not only resulting in high market shares but also allowing them to (at least temporarily) reap above-normal profits as a result of market power. This argumentation is also empirically supported. Some empirical evidence, for example, comes from a study by Hendricks and Singhal (1997), cited in Nehausler et al. (2011), who showed that delaying the introduction of new products decreases the market value of the firm. Certainly, significant penalties for firms seem to exist for not introducing new products on time.

Press releases, company statements, and secondary texts and articles clearly indicate that China Mobile, Huawei and ZTE all clearly understand the importance of market speed and the need to engage in intellectual property rights protection in order to secure market share and capture the benefits of the emerging 4G ecosystem. For example, in an article entitled "The Great Leap Forward: How the World's Largest Operator Aims to Jump One Generation", China Mobile's research unit's general manager Bill Huang asserts: "telecom vendors are not moving fast enough to develop products for the 4G ecosystem" (Huang cited in Thoren and Harvard, 2011). Moreover, Huawei's Vice President of Wireless Marketing Lars Bondelind recently asserted that the company possesses a three- to six-month lead over its closest rivals, Ericsson and Nokia Siemens (NSN): "Let's put it this way; we have a cost advantage because of our R&D base in China, which is 35,000–40,000 engineers. That is a big advantage. ... But, today, we are three-to-six months ahead of most of our competitors in all areas. We are competing with Ericsson and NSN on features and functions and technology. And they are playing catch-up" (Bondelind, cited in Sun, 2011). In a similar vein, ZTE has articulated its intention to become one of the top three global wireless equipment providers globally. As Xu Ming, vice-president of wireless services for ZTE, told *Reuters*: "We want to be in the top three in terms of revenues and market share". Xu also stated that international expansion was a major priority for the company:

> "Our strategy for the wireless business in the past 10 years has been to gain critical mass in China, and then we expanded into emerging markets like Africa, Asia, and Latin America. Now we are bringing that push into developed markets like Europe" ... Xu stated that he was confident that ZTE could make its goal of reaching the top three. The trend is very clear. If you look at Alcatel Lucent or Nokia Siemens Networks, their growth rate is flat or shrinking, and even market leader Ericsson has a slow growth rate. For ZTE, if we can continue to grow at the very rapid rates that we have seen in recent years, we will soon take over one of the major vendors in terms of revenues. (Xu, cited in Weissberger, 2011)

China Mobile is one of the most aggressive operators when it comes to pushing for the commercialisation of TD-LTE technology. With the approval of MIIT to conduct large-scale testing of TD-LTE technology, China Mobile has tapped vendors such as Ericsson, Alcatel-Lucent and Nokia Siemens to deploy various pieces of the network in six large Chinese cities. The six cities are: Shanghai, Hangzhou, Nanjing, Guangzhou, Shenzhen and Xiamen. China Mobile said in a statement that the upcoming large-scale tests are aimed at exploring the commercial potential of the technology, and to provide impetus for international telecom carriers to adopt and deploy the new network. As the chairman of China Mobile recently conjectured: "We are targeting commercialisation next year, not in five years......So you see: 4G is not being pushed by the vendors, like 3G was. 4G is being pushed by the carriers. LTE is the only standard in the industry where, if you have a product, people will buy it right away. It's the reverse of how things used to be, and very interesting. LTE is being developed fast, but not fast enough" (Huang, cited in Thoren and Harvard, 2011). It has been estimated that the market for LT-TDD handsets, equipment and semi-conductors will total $98 billion from 2012 to 2016 (Goldman-Sachs, 2011).

Indeed, TD-LTE is currently being widely adopted by the vendor community as well as operators. According to Thorsten Robrecht, head of LTE Radio Access Product Management at Nokia Siemens Networks, TD-LTE has emerged as a solution with a far wider range of potential applications than was first expected. "'TD-LTE is really a global LTE solution, I'm very sure about this,' he says. 'It has evolved so much over recent months and now I'm seeing it appearing in all sorts of countries.' While the standard has been incubated in China, Robrecht says, the earliest deployments are more likely to come from elsewhere. He reports significant activity in Japan as well as India and Russia – and even some customers considering it in Australia" (Nokia Siemens, 2011).

Moreover, whilst China Mobile – and hence, the Chinese government – is certainly the primary operator driving TD-LTE, the technology has now achieved global momentum:

> China Mobile with its half a billion subscribers can clearly define a market segment on its own. However, it is the interest from other major markets – such as Russia, Japan, India and the US – that has put TD-LTE on every operator's plan....Currently, overseas operators from [a] dozen countries and regions in Europe, America and Asia have begun their TD-LTE cooperation with China, and many of them plan to start trial network construction in 2010 and even to deploy commercial networks. (C114, 2010)

For example, in June 2010, Qualcomm won India's 2.3 GHz Broadband Wireless Access (BWA) spectrum auction for TDD development in four regions of the country. Furthermore, Qualcomm has promised to set up a JV to build an LTE network. This indicates that Qualcomm has TD-LTE in its strategic development plan. Mobile operators in the US, including AT&T, Verizon and America Mobil, have also promised to support TD-LTE. Leading operators in Japan and Korea have also played an active role in promoting TD-LTE. Other operators with WiMAX, Personal Handyphone System PHS (1900–1920 MHz), and TD-SCDMA will also choose TD-LTE as their migration path (Min, 2011).

Consequently, it can be argued that China Mobile is becoming a global leader in the sector and has acquired the necessary networking capabilities and relational legitimacy to play a key role in shaping the TD-LTE ecosystem. As it has already been established, system shapers and ecosystem coordinators play a key role in the contemporary socio-technological system by ensuring system convergence and interoperability via the coordination of large sets of system actors. The overarching result here is that the state itself is linked to the technological ecosystem via a dense array of deterritorialised or partially deterritorialised alliances with public and private-sector actors. As China Mobile's (former) Chairman Wang Jianzhou, asserted: "TD-LTE technology has opened up a 'global market' for the company. The next important step for China Mobile is to converge TD-LTE technology with the other mainstream 4G technology – LTE FDD. ... Convergence will provide a much bigger platform for TD-LTE technology" (Wang, 2011).

Both Huawei and ZTE have also been successful in securing significant market share in the global wireless communication sector in recent years. For example, both Huawei and ZTE were ranked in the top five telecommunications vendors in the world (by revenue) in 2011: Huawei took second place with $15.4 billion and ZTE took fifth place with $5.8 billion (OECD Information Technology Database, 2011).

Huawei has customers in 130 countries and sells equipment to 45 of the world's top 50 telecom operators, and it had $28 billion in revenue in 2011. In the same year, it ranked 352 on *Fortune*'s Global 500 list. Huawei has continued to increase its market share in global markets. For example, Huawei was awarded 36% of LTE kit tenders in 2010 (Mobile News, 2010). This is estimated to be twice the number so far secured by Ericsson at 16% (Mobile News, 2010). NSN also acquired 16%, Alcatel-Lucent 14%, and others 18%. With sales projected to grow another 10% this year, it has been suggested that Huawei will soon surpass Sweden's

Ericsson and assume the position as the world's No. 1 manufacturer of communications equipment (Mobile News, 2010).

It is important to note here that this increase in market share is not just derived from domestic sales. For example, in 2012, Huawei had total sales of 220,198 yuan. Of this 73,579 yuan was derived from the Chinese domestic market, 37,359 34 yuan from Asia Pacific, 31,846 yuan from America and 77,414 yuan from Europe, the Middle East and Africa (EMEA) (Huawei, 2012). Hence, it can be seen that a significant proportion of Huawei's market share was derived from increasing global market share as opposed to domestic sources.

ZTE is also experiencing significant growth in the LTE market. In the first half of 2011, ZTE's new LTE contracts exceeded its total number of LTE contracts for all of 2010. For example, ZTE won 28 contracts for LTE commercial applications and deployed test networks in cooperation with more than 90 operators across the globe. To date, over 100,000 ZTE LTE terminal units have been ordered worldwide. The terminals, which are purchased by global high-end operators, contribute to the commercialisation of LTE technology (Min, 2011).

Moreover, in 2012, ZTE reported that 39.56 billion yuan of total sales revenue was derived from the domestic market, accounting for 47% of total sales whilst 44.66 billion yuan was derived from the international market, accounting for 53% of the group's overall sales revenue (ZTE, 2012). Conversely, it can be seen that ZTE, like Huawei, is deriving more sales revenue from the global market than the domestic market.

Furthermore, at the end of September, ZTE and Huawei both announced that they would build the TD-LTE network for Japan's SoftBank. This network will cover 90% of the population in Japan and will be the largest TD-LTE project across the world (C114, 2011). Before this project, ZTE and Huawei built the first large-scale LTE-TDD/FDD commercial network together for Hi3G. In October 2011, ZTE and Huawei also announced that they will build India's first TD-LTE commercial network for Bharti Airtel, India's largest mobile carrier. Moreover, it has been estimated that ZTE will account for 35% of TD-LTE equipment supplies worldwide. The forecast released by Goldman Sachs indicates that the total global TD-LTE CAPEX could reach US$15 billion to US$20 billion from 2012–2014 with more than 40 million subscribers (Goldman-Sachs, 2011). Conversely, it can be seen that these firms are securing significant market share in the 4G sector and are firmly embedded in both the 4G ecosystem R&D process as well as being at the forefront of its commercialisation rollout.

Thus, it can be argued that China Mobile's adoption of a more collaborative business model that is focused around the idea of ecosystem and network building at a global level has proved to be an effective developmental model that has helped to embed it in the 4G market infrastructure at an early stage of the sector's development. Furthermore, these recent cooperative linkages via strategic alliances in the Chinese wireless communication sector suggest that the Chinese government is moving beyond its policy of building national domestic champions to one of state-led alliance capitalism, where the key policy goals are designed to facilitate global cooperation, ecosystem alliance building and the co-development and trans-territorial appropriation of technological assets and knowledge.

State-led alliance capitalism from this perspective represents a strategic state response by the Chinese government to utilise recent changes in the contemporary socio-technological environment in order to enhance its comparative advantage in high-technology markets. Specifically, it is a trans-territorial developmental approach that addresses the critical ecosystem dependencies via collaborative R&D and regulatory decision-making processes at an early stage in a technology's development. As this book has sought to highlight, in the contemporary global wireless communication sector, the development of standards, intellectual property assets and compatible ecosystem products begins at an early stage in the developmental process and is increasingly defined by alliance-based collaborative processes in order to ensure system convergence and interoperability. For the Chinese developmental state in the global wireless communication sector this involved playing an early role in the standardisation process and facilitating the recruitment of alliance members for the 4G TD-LTE ecosystem.

In order to achieve its high-technology developmental goals, the Chinese state leveraged the power of network coordination and its increasing global relational and regulatory capacity in order to motivate a large number of system participants to make significant investments in the 4G TD-LTE technology and to accelerate the development of the ecosystem itself and the associated commercialisation processes. In this way, new techno-economic alliances have been made between the Chinese developmental state and foreign firms, and important technological, economic and political linkages to the global economy have been forged.

This has not been a simple linear process. Instead it has involved a process of policy experimentation and adaptive learning. This can be seen manifest in the way in which the Chinese government has continuously adjusted its policy prescriptions in response to socio-technological system barriers and opportunities in the wireless technology sector. In this way, the Chinese state, as a set of bounded rational actors, has engaged in a new round of capability building, new policy tools and methods of strategic, global, economic diplomacy that are highly responsive to socio-technological systems change. In many ways the strategic agency employed by the Chinese state corresponds to the behaviour exhibited by global network states. As argued in Chapter 4, global network states endeavour to shape the capabilities of the business sector and society itself and its associated markets (O'Riain, 2000:166). This involves the building of local and global network capabilities and innovation systems. However, whilst the GNS actively engages in technological and social systems shaping, the interventionist Chinese state and its unique form of state capitalism has gone a step further and has been actively engaged, via state-owned China Mobile, not just in the shaping of the socio-technological system but in the construction of the technological ecosystem itself and its associated regulatory environment. This form of developmental state is defined not just by its system-shaping abilities but by its ability to acquire significant ecosystem power and regulatory capacity in the contemporary socio-technological system.

Chinese government support and technological development in the global wireless communication sector

It is important to emphasise at this point that an assessment of Huawei and ZTE's patenting activity and increasing market share correlates with governmental increases in R&D funding, tax concessions and S&T policy directives. Specifically, the 863, 973, and Torch programs that were initiated in the late 1980s and 1990s, and the Medium to Long-term Plans for Science and Technology first initiated in 2006, are all key governmental policy programs designed to accelerate the S&T development of the nation. Furthermore, the fact that the telecommunications sector in which Huawei and ZTE operate was declared by the Chinese government to be strategically important has afforded promising companies operating in the sector significant governmental support. Indeed, since the mid-1980s, the government's industrial policy in the sector has facilitated a transformation of the network from a creaking, thinly spread system to a world-class utility that rivals some of the world's most advanced countries (Harwit, 2007).

For example, before 2001 Huawei and ZTE engaged in little patent activity. From 2002, this begins to change with moderate amounts of patent activity being conducted by Huawei and ZTE. Moreover, it is not until 2004–05 that we see a sharp increase in the patenting activity of both firms. For example, between 1985 and 2006, ZTE filed a total of 61 patents. Within this period, Huawei filed 399 patents (Eberhardt, Helmers and Zhihong, 2011:9). By 2008, Huawei has increased its patent activity significantly and moved to be the world leader in the PCT's patent applicants ranking. In a similar vein, by 2011, ZTE moved to first place in this ranking, while Huawei dropped slightly to third (Godinho and Ferreira, 2013:1044).

It can be conjectured that this increase in patenting activity reflects the Chinese government's commitment in the period to accelerate the science and technology development process. For example, since 1998, the Chinese government tripled its financial support for higher education and universities committed to basic and applied research. As a direct result, the number of students quintupled from 1 million in 1997 to 5.5 million in 2007. Moreover, since the early 1990s, China's domestic science and engineering doctorate awards have increased tenfold to approximately 21,000 in 2006 (Ernst, 2010). What is important to emphasise is that this increase in highly trained students provided both Huawei and ZTE with a labour force capable of engaging in the development of patents and products for the global wireless communication sector. Indeed, the capacity to innovate is directly tied to the quality of the labour force. As Toner (2011) asserts: "When more skilled labor exits, the market for skill-complementary technologies is larger. More of these technologies will thus be invented and they will be complementary to skills promoting faster upgrading of the productivity of skill workers". That is: "an increase in the supply of skills can generate skill biased technical change" (Toner, 2011).

Thus, it can be argued that the Chinese government's strategic industrial policy agenda in the sector has played an important role in increasing the technological capacities and the internationalisation process of Chinese telecommunications companies and hence their increasing patent numbers and global market share. This assertion, it is argued, can be substantiated by the correlations that exist between firms such as Huawei and ZTE's increasing global success and innovative capacity and the industrial policy directives issued by the Chinese government that have defined the sector for the past 25 years.

Political implications of the new program

The globalised use of "strategic agency" and "ecosystem shaping" by the Chinese government has a number of significant political and global implications. Firstly, it inserts the Chinese government directly into the global 4G wireless communication sector's developmental processes as active technological and developmental participants, as opposed to a nation engaged in technological catch-up via FDI and technology transfers. This confers on it significantly more globalised market and technological power than it had in the earlier stages of the development of wireless technologies. It owns key patents, has members of international patents committees and developmental boards, and has actively engaged in shaping the 4G ecosystem. Secondly, it signifies the transformation of the relationship between the Chinese government and foreign firms as new sets of organisational and trans-territorial collaborative relationships evolve. As I have highlighted in this book, this has involved a raft of new legislation and programs designed to allow foreign businesses operating in the sector to compete on equal terms with politically well-connected domestic businesses as long as they can provide technological value to the wireless ecosystem in a way that benefits Chinese technological development interests.

The success of the Chinese government in the 4G sector, it can be conjectured, provides support for the idea of state capitalism as a viable political and economic model for technological upscaling in the twenty-first century – especially in sectors where R&D has long time horizons and requires significant funding. For the Chinese state this has provided both a source of domestic pride and political legitimacy. Their new form of hybrid state capitalism bridges the need for governmental funding and supportive developmental policies with the market in new and important ways that does not act to pick winners per se, but to co-facilitate and mould the technological developmental process in a cooperative fashion at multiple levels. The overarching result is that Chinese popular support for the idea of state capitalism is likely to increase and the role of the state in economic and technological development in China is likely to remain strong in the near future.

Conclusion

In this chapter, I have sought to highlight how the Chinese government and its leading domestic firms have adjusted to the necessity of global technological embeddedness and the innovation imperatives that define the global wireless communication sector. I have illustrated how new forms of collaborative relationships between Chinese domestic and foreign firms, multilateral institutions and the Chinese government in the sector are conspiring to expand the territorial scope and behaviour of both the firm and the state. From this theoretical perspective, alliance capitalism, as both a form of global economic organisation and embedded structural and relational interdependence, is facilitating the institutional transformation and behaviour of all actors in the sector, as they are compelled to collaborate at a global level to develop core technologies and embed themselves in the global wireless communication sector's ecosystem and commercialisation process.

Furthermore, in this chapter I have illustrated how the interests of Chinese firms and TNCs have become structurally and relationally interlinked in a way that affords the Chinese government and its leading domestic firms more bargaining power and increased rights and patterns of participation in the sector. Indeed, the capacity of the government of China to respond to foreign concerns in a way that has allowed them to develop its domestic wireless communication sector, whilst also satisfying the interests of the many actors involved, presents us with an interesting case of "international political and economic bargaining and policy adjustment" as well as a frame of reference to examine the actual processes associated with Chinese policy making in a globalised economic context.

7
Conclusion

In this book I have sought to highlight how socio-technological systems change and the advent of a more collaborative economy in the global wireless communication sector has created new sets of system pressures and developmental opportunities for the Chinese government and its domestic high-technology firms operating in this sector. Specifically, I have argued that the concepts of "convergence" and "interoperability" are key structural change drivers that have facilitated the development of new sets of critical ecosystem dependencies that require technological actors to engage in collaborative alliance-based behaviour in order to ensure technological ecosystem convergence and facilitate the development of high-technology assets and products.

These alliances, it has been argued, are not just designed to appropriate complementary assets and resources as earlier alliance forms did; instead, they are designed to facilitate the development of other key ecosystem actors in a bid to ensure that critical infrastructure, regulatory standards and complementary technologies are developed before a technology goes to market. I have termed these new alliance forms "innovation ecosystem-building alliances". The primary goal of such alliances, it has been asserted, is to anticipate future market and ecosystem requirements and to utilise this information in order to construct a critical network of interdependent alliance partners that are focused on achieving techno-logical convergence and interoperability in the sector.

I have also highlighted how the Chinese developmental state has adopted a highly adaptive globalised response to the emergence of this more collab-orative, alliance-based, socio-technological system. The empirical focus has been on the 3G and 4G wireless technology sector. I have argued that after the failure of FDI and techno-nationalism as developmental strate-gies for the development of 3G technologies, the Chinese government has undergone a shift in its technological development agenda, R&D allocations and regulatory and relational policy tools. Essentially, in order to participate in the development of the 4G technological ecosystem, it has endeavoured to forge and reinforce R&D, organisational and regulatory alliances with foreign firms and global regulatory actors. Furthermore, I have argued that this state-led, alliance-based, developmental strategy in the 4G technology sector has been highly successful and has resulted in the favourable insertion of Chinese domestic firms, and hence, China's economy, into global innovation and production networks in the sector.

Empirical findings

The empirical findings of this book support the key theoretical assertions pertaining to the emergence of alliance capitalism in the global wireless communication sector and the ability of the Chinese government and its key domestic firms to employ "strategic agency" and "ecosystem-shaping" policy tools in order to successfully ensure both the technological upscaling of Chinese domestic firms operating in the sector and their embeddedness in key wireless technology ecosystems globally. Firstly, empirical data directly related to China Mobile, specifically, China Mobile's role in the development of a 4G TD-LTE ecosystem, and its associated alliance structures via the development of anticipatory standards and the setting-up of the Global TD-LTE Initiative (GTI), clearly indicates that China Mobile has been highly successful in facilitating the construction of a global collaborative platform for the co-development of TDD and FDD standards and the development of a trans-territorial ecosystem for the organisation of technological coordination and cooperation around the standards at a global level. Evidence that China Mobile has successfully positioned itself as a key network organiser and socio-technological systems shaper in the global wireless communication sector is also supported by the large number of firms willing to join the ecosystem and develop the necessary technologies and products required for system convergence and interoperability.

Secondly, as this book highlighted in Chapter 6, R&D, patent, alliance and market-share data indicate that Chinese domestic firms are highly adaptive fast followers that are achieving significant global market share in the 4G wireless sector as both ecosystem builders and network organisers via the development of new technological platforms, markets and collaborative alliance structures.

Thirdly, this book has provided empirical data to support the assertion that the Chinese developmental state operating in the contemporary wireless technology sector is engaged not just in fixing market failures, but also in the active creation of markets for the new technologies via the provision of an opportunity space for both foreign and domestic firms. This has involved the Chinese state providing trans-territorial development funding and governance based on ecosystem development needs rather than domestic capacity enhancement or national pillar innovation goals. These changes in sectoral funding and the trans-territorial extension of Chinese governance in the sector via standards-making bodies and globalised technological development networks and knowledge zones indicate that socio-technological systems change is indeed impacting on the strategic planning and agency of the Chinese government, its key domestic firms and international ecosystem partners. Specifically, this book has highlighted how Chinese policymakers in the sector are responding to this socio-technological systems change by developing important sets of new state policy tools and capabilities such as ecosystem shaping, network organisation, technological co-development, trans-territorial development funding and system embeddedness.

Theoretical review and implications

The central theoretical focus of this book has been on the emergence of alliance capitalism as an important new form of economic organisation with a specific theoretical focus on the role of the Chinese developmental state in both facilitating and governing the parameters of global technological alliances in the wireless communication sector. It has been argued that alliance capitalism, as a new form of global economic organisation and a form of embedded structural and relational interdependence is encouraging the institutional transformation and behaviour of all actors in the sector, as they collaborate to develop core technologies and embed themselves in the global wireless communication sector's ecosystem and commercialisation processes. Certainly, the fact that technological innovation and product development in the contemporary knowledge economy occur in globally and organisationally dispersed innovation networks has significant implications for Chinese technology developmental policy and governance structures. What needs to be emphasised here is that because these technologies are created and developed in contemporary innovation networks that are increasingly defined by economic space instead of political geography they partly transcend national or geographical governance structures.

Hence, as this book has highlighted, in order to effectively operate in this changing socio-technological environment, the government of China has initiated a strategic shift away from ideas of indigenous technological development and national sectoral pillars to a socio-technological development strategy that is informed by the concept of "globalised adaptive ecology". "Globalised adaptive ecology" has been used in this book to refer to strategic policy responses that combine both state agency/intervention and developmental planning with the need to gather, co-create and appropriate technological knowledge and assets from multiple states and technological knowledge zones.

In this book, I have argued that the idea of globalised adaptive ecology and structural socio-technological system shaping is an important theoretical medium from which to analyse the adaptive behaviour of the Chinese government and Chinese firms, and their ability to operate and compete in multiple technological domains and system levels. Specifically, I have conjectured that key analytical concepts, such as co-evolution, alliance capitalism, self-organisation, complex adaptive ecology, dynamic structural change, knowledge coordination, experimental trial-and-error learning, strategic system shaping, actor agency and trans-territorial multilevel governance, all need to be incorporated into contemporary theoretical analyses.

I have also conjectured that in order to understand exactly how technology facilitates and shapes the emergence of socio-technological systems and how these large-scale global technological and knowledge systems are being politically, economically, socially and scientifically shaped and organised at the national, global and sub-national levels, we need to comprehend how the system itself functions and the key variables that encapsulate its evolution, development and governance. Theoretically, this requires us to examine and frame our analyses with a framework capable of understanding how complex adaptive systems contribute to technological evolution and socio-technological systems development.

Furthermore, as this book has highlighted, competition in this emerging environment has shifted from one that is focused on intellectual property monopolisation and market power to one of ecosystem membership and embeddedness. This is because in order to participate in the global networks and sectoral systems that make up the contemporary socio-technological system in the wireless communication sector, actors need to develop technological innovations that possess the ability to converge or interact with other firms. Furthermore, the ability to ensure convergence and interoperability is highly dependent on the capacity of firms to access globally dispersed scientific and technical knowledge sources and to generate long-term, networked relational ties with other system actors.

The Chinese state has played a key focal role in the coordination of such collaborative processes as a network organiser and system integrator. By helping to both facilitate the insertion of its domestic firms into the 4G global innovation ecosystem and by also actively constructing a niche technological ecosystem by developing its own technological standard (4G TD-LTE) that could converge with other standards in the sector and recruiting members to populate and develop technological assets for the new ecosystem, the Chinese government has, for the first time, acted not just as a system shaper but also as a system maker. It has done this by integrating international firms into its domestic economy and innovation platforms, and by offering a range of incentives designed to encourage the development of a complex network of extra-territorial linkages that could facilitate the development of globalised technological niches and ensure the interoperability and openness of key emerging innovation platforms and their associated technologies. In this way, the Chinese state has ensured the technological advancement of its domestic firms as well as the ability to appropriate technological rents from multiple globalised sources. The emergence of collaborative global innovation in key high-technology sectors means that very different policy capacities are required compared to when policy objectives were solely centred on the idea of indigenous technology and the idea of building national champions. This is because the state's territorial jurisdiction over knowledge is becoming less important than its capacity to access and organise complex networks of collaborative and dispersed innovation processes. Thus, relational and globalised regulatory capacities have become important strategic policy tools that the Chinese developmental state has utilised in order to facilitate technological alliances between foreign and domestic firms and to appropriate value from emerging high-technology ecosystems.

In this dynamic and highly competitive environment, development strategy is no longer a matter of positioning a fixed set of activities along a neoclassical industrial model, the value chain. Instead, successful companies and state actors are endeavouring not just to add value but also to co-develop and reinvent value. For the state, its role in the creation of economic growth is becoming increasingly dependent on its ability create an ecosystem of firms that transcend the territorial boundaries of the state. Essentially, for the Chinese developmental state, it is essentially a "third way" whereby the states strategic control over the globalisation process is modified in a way that still allows it a certain degree of control over the developmental process itself, but in which the existence of critical globalised ecosystem dependencies and the interests of global firms are actually integrated into the Chinese developmental strategy. Hence, the goal of successful high-technology governance in a contemporary context is not to control the socio-technological environment and appropriate nationally controlled intellectual property assets or to focus on the protection and development of nationally developed sectors. Rather it is to enable and adapt, and partially shape via collaborative coordination the emerging socio-technological environment. Conversely, the ability to self-organise, adapt and develop generative extra-territorial relationships is an important system capacity required by the current socio-technological system.

In order to achieve these key developmental goals, new state-interventionist policy tools have emerged that are designed to address the increasing need for the state to play a role in funding long-term risk capital to help facilitate the development of technologies with long time frames in high-technology sectors such as the global wireless communication sector. From this theoretical perspective, state intervention is a key prerequisite for successful high-technology upscaling. Indeed, for the Chinese developmental state operating in the global wireless communication sector, strategic agency has been derived from its ability to engage in a new round of capability building and develop new policy tools and methods of strategic global engagement that are highly responsive to socio-technological systems change. I have argued in this book that the strategic agency employed by the Chinese state corresponds to the strategic behaviour exhibited by global network states, which have emerged in response to a more collaborative global economy and primarily endeavour to link domestic and foreign actors together in a way that facilitates the co-production and appropriation of technological assets (O'Riain, 2000: 166). The GNS developmental state, asserts Sean O'Riain, is defined precisely by "its ability to create and animate Post-Fordist networks of production and innovation and international networks of capital and to link them together in ways which promote local and national development" (O'Riain, 2000:166). This requires a fundamental shift in government policy directives, from commanding specific outcomes to creating and maintaining new markets, and a move away from top-down policy development to steering and negotiating its intentions with partners in both the domestic private sector and the international sphere (Jessop, 1997:96).

In accordance with the central theoretical precepts expounded by the GNS, the Chinese state has targeted strategic investments to facilitate research and development and alliance participation between its domestic and foreign firms. In this manner the Chinese state has sought to build both local and global network capabilities. However, the Chinese developmental state differs in a number of ways from the GNS. Firstly, it has endeavoured to enhance its strategic agency by not just actively engaging in technological and social systems shaping, but also it has sought to use its state-owned firms to facilitate the construction of the 4G TD-LTE technological ecosystem and its associated regulatory environment. I have argued that this form of developmental state is defined not just by its system-shaping abilities but also by its ability to acquire significant ecosystem power and regulatory capacity in the contemporary socio-technological system. Indeed, the Chinese developmental state provides us with an interesting model of a state-led developmental strategy that poses a serious challenge to technological and economic discourse as it is currently framed. State-led alliance capitalism certainly presents a challenge to the developmental agenda as defined by Western nations, offering a developmental path to be emulated by other developing nations. It provides us with an interesting and important model for examining how technological and economic power relations will play out in the near future and forces us to conceptualise markets in a more dynamic way, incorporating into our analysis new configurations of ecosystem power and interests and models of technological development.

However, current theoretical approaches used to understand Chinese high-technology development are primarily framed around China's latecomer status and failed techno-nationalistic policy program and do not take into account recent policy modifications and socio-technological adjustment strategies being devised by the Chinese government. Specifically, the literature fails to adequately address the way in which recent policy directives are increasingly framed by an acknowledgement by Chinese government leaders – that the way in which global innovation works and the need to address critical ecosystem dependencies requires a shift towards a more trans-territorial collaborative socio-technological developmental plan. Indeed, alliance capitalism and the co-production of technological assets that it entails has fundamental implications – not just for the Chinese state and its domestic firms, but also for the global high-technology system and its system of intellectual property rights, relational networks and modes of value appropriation. Hence, researchers need to begin to explore exactly what types of relational and regulatory capacities are emerging? What does this mean for intellectual property rights and value appropriations measures? What role does the state and its SOEs have in advancing not just the high-technology upscaling of its own national firms but the upscaling of foreign firms when the primary focus has shifted from one of building national champions to one of trans-territorial technological ecosystem development?

Appendices

Appendix A – Chinese firm questions

1 It is well known that foreign companies manage virtually all intellectual property and account for 85% of the country's technology exports. Do you feel that "Wintelism", as it has been termed, is inhibiting the technological advancement of your firm in the wireless communication sector, specifically in the development of 3G and 4G technologies?

2 Do you feel that your firm has developed a dependence on transnational capital and technology?

3 Is your firm currently engaged in any attempts to develop indigenous standards and technology platforms in the 3G and 4G wireless communication sector?

4 Does your firm engage in horizontal collaboration with domestic counterparts in the wireless communication sector?

5 How important is it for your firm to forge collaborative relationships with Beijing and its regulatory institutions in order ensure the successful functioning and development of your firm?

6 Are you currently engaged in any alliance or collaborative activity with a foreign firm?

7 If so, how many foreign firms are you collaborating with? What is the exact nature of these relationships and are these collaborations short or long term in duration?

8 What is the exact nature of these collaborations, and do you expect these collaborative activities to result in your firm assuming either sole or joint ownership over proprietorial intellectual property in the 3G or 4G wireless communication sector?

9 How strategically important is it for your firm to establish long-term relational assets and network connections globally in order to embed itself in global R&D networks in the wireless communication sector? That is, does your firm consider relational assets, global network ties and collaborative capacity to be important to its future developmental success?

10 Could you please give me your understanding of a strategic alliance? What do you consider to be the most important assets that your firm can gain when engaged in a strategic alliance with a Chinese firm?

11 Do you feel that your alignment with foreign firms and the need to develop a global market strategy and to gain access to codified technology conflict with the strategic developmental goals and interests articulated by Beijing?

12 Do you feel that Beijing's attempt to define the boundaries of the wireless communication sector's market construction and market access has served to define the overarching path of your company's development or that you have a degree of interdependence in defining the parameters of your firm's development strategy?

13 Are you members of any transnational business alliance? If so, what are your objectives for joining this type of alliance, and what do they entail?

14 How much of your firm's budget is devoted to R&D?

15 What role does the development of anticipatory standards and pre-competitive market construction play in the strategic development of your firm's technological assets? That is, what percentage of your firm's budget is allocated to "primary", "radical" or "first generation" R&D?

16 It has been suggested that the concept of R&D is being replaced by the idea of concept and development (C&D); that is, "connect and develop". How much of your current budget is allocated to C&D?

17 It has recently ben asserted that due to the complexity of the knowledge economy and the need for collaborative R&D and product development at a global level the "network" is replacing the firm as the dominant actor in the globalised, knowledge-based economy. Do you agree or disagree with this assertion?

18 Do you think that a networked global knowledge economy will/is resulting in the development of reciprocal interdependence between foreign and domestic firms at a global level?

Appendix B – Foreign firm questions

1 It is well known that current global innovation networks are controlled by lead firms ("flagships") primarily from the West that dominate control over network resources, technology and decision making. How do you think this economic and technological concentration impacts on Chinese firms wishing to develop an indigenous innovative technology base?

2 The recent Indigenous Innovation Product Accreditation Program indicates that Beijing is attempting to redirect FDI in the manufacturing sector towards the R&D sector and facilitate competition

and cooperation between Chinese and foreign firms. Do you think these policy directives represent an attempt by Beijing to embed technologically advanced foreign firms into the Chinese economy?

3   How do you feel these strategic changes will impact on your firm? Do you think Beijing has formulated a "calculative" macro-strategy and is using the size of its domestic market as a strategic tool to pressure foreign firms to transfer cutting-edge technologies to Chinese firms, which can then use them to complement and accelerate national economic development?

4   Do you believe that exchanging technology for market access in China poses significant risks for foreign firms operating in China? That is, are you creating future competitors? Could you please outline these risks and any strategies that you utilise or that could be utilised to minimise this problem now and in the future?

5   Beijing has recently announced its intentions to pursue a leading role in the development and marketisation of 4G wireless technology. In order to achieve this, the country has recently developed its own technological standards such as TD-SCDMA. It has also launched a number of government-sponsored research projects on 4G. Do you feel obligated to adopt Chinese-made standards and develop new product lines compatible or interoperable with new or expected PRC standards in order to ensure future access to China's domestic market?

6   As you are probably aware, Beijing's attempts to appropriate the lock-in effects associated with technological standards ownership and the comparative advantages that it is thought to confer has met with fierce international resistance. As a result, Beijing has sought to modify its policy framework with one that is more receptive to a more open-market focus and which emphasises the need to meet the requirements of the international regulatory environment. Within the wireless communication sector, this new policy framework focuses on the development of pre-competitive R&D, anticipatory standards and technological alliances with foreign firms. Zhang Yansheng, director of the Institute for International Economic Research under the National Development and Reform Commission (NDRC), has asserted that Beijing prefers strategic alliance multinationals willing to establish R&D centres in China.

7   Do you currently have R&D centres in China?

8   Could you please outline the organisational, managerial and key goals of these R&D centres?

9 How does the management of your R&D facilities in China differ from those at home? Do you feel obligated to set them up or is it a necessary part of your long-term plan for accessing and gaining share of the Chinese market via the adaption of products for the local market?

10 If you do not currently have R&D centres in China, do you have plans to open R&D centres in China in the near- or long-term future?

11 Are you willing to share a greater proportion of your firm's profits, technological knowledge, organisational structures and network connections with Chinese firms now or in the future?

12 Could you please give me your understanding of a strategic alliance? What do you consider to be the most important assets that your firm can gain when engaged in a strategic alliance with a Chinese firm?

13 How strategically important is it for your firm to establish long-term relational assets and network connections globally in order to embed itself in global R&D and innovation networks in the wireless communication sector? That is, does your firm consider relational assets and collaborative capacity to be important to its future developmental success within the Chinese domestic market globally?

14 What role does the development of anticipatory standards and pre-competitive market construction play in the strategic development of your firm's technological assets, specifically those related to 3G and 4G wireless technology?

15 How much of your firm's budget is devoted to R&D designed to develop the 3G and 4G technology sector?

16 How much of your firm's budget is devoted to collaborative R&D with Chinese firms working to develop 3G and 4G wireless technology?

17 It has been suggested that the concept of R&D is being replaced by the idea of concept and development (C&D); that is, "connect and develop". How much of your current budget is allocated to C&D?

18 Does the networked nature of the contemporary global economy mean that all firms will need to work together in a more collaborative fashion to construct and share wireless technology and its future standards in order to ensure seamless interoperability in the sector?

19 It is clear that Western markets are highly saturated and driven by replacement demand and, as a direct result, the Chinese domestic market has significant market potential for foreign firms. What are your key motives/goals/objectives for entering into an alliance with

Chinese state-owned or private firms in the wireless communication sector?

20 How important is it for you as a foreign firm in China to forge collaborative relationships with the Chinese state and its regulatory institutions in order to ensure the successful functioning of your firm in China?

21 How does your interaction with a state-owned Chinese firm differ from that of a privately owned firm such as Huawei?

22 What strategies or avenues do you use when bargaining and negotiating terms of access with the Chines government?

23 Are you a member of any transnational business alliance? If so, what are your objectives for joining this type of alliance and what does it entail?

24 The last decade has witnessed substantial growth in the talent pools of emerging economies – especially in Asia. Because salaries make up a significant proportion of the cost of R&D, is the availability of highly skilled human resources at lower cost an important factor in your decision to relocate your R&D in China?

25 It has recently been asserted that due to the complexity of the knowledge economy and the need for collaborative R&D and product development at a global level, the "network" is replacing the firm as the dominant actor in the globalised, knowledge-based economy. Do you agree or disagree with this assertion?

26 Do you think that a networked global knowledge economy will or is resulting in the development of reciprocal interdependence between foreign and domestic firms at a global level?

Appendix C – Government/ministerial/academic questions

1 Recent policy directives such as those contained in the 2006, 15-year "Medium to Long Term Plan for the Development of Science and Technology" and the Indigenous Innovation Product Accreditation Program indicate that Beijing wants to develop an innovative indigenous knowledge economy. What is your understanding of Beijing's key development objectives here?

2 It is well known that current global innovation networks are controlled by lead firms ("flagships") primarily from the West that dominate control over network resources and decision making. How do you think this economic and technological concentration impacts on Chinese firms wishing to develop an indigenous innovative technology base?

3 The recent Indigenous innovation Product Accreditation Program indicates that Beijing is attempting to redirect FDI in the manufacturing sector towards the R&D sector and facilitate competition and cooperation between Chinese and foreign firms. Do you think these policy directives represent an attempt by Beijing to embed technological advanced foreign firms into the Chinese economy?

4 The Indigenous innovation Product Accreditation Program contained some rather stringent policy directives (the government has now modified these requirements due to international political lobbying). For example, foreign companies patent technologies in China and adopt Chinese standards if they want to sell in the Chinese market. Another provision requires companies to pay Chinese employees at least 2% of profit derived from their inventions in China unless the employees explicitly waive that right. This emphasis on indigenous innovation and what some are calling forced technology transfer has raised concerns over the rise of "techno-nationalism" or "neo-techno-nationalism" in China and the future economic openness of China as well as the protection of foreign intellectual property in China. Do you think these concerns are legitimate?

5 The international response to these plans was swift and coordinated, and Beijing has now modified these requirements due to international political lobbying. Has Beijing been forced to become more receptive to the needs of international regulatory regimes and global economic actors in order to participate in the emerging global knowledge economy?

6 If so, how exactly has this impacted on Beijing's indigenous development plans?

7 It has been argued that the network is replacing the firm as the dominant actor in the increasingly globalised, knowledge-based economy. It has also been suggested that in order to build an innovative, technologically advanced, knowledge economy, and due to the globalised nature of contemporary R&D, that few if any products are developed in a single territory, and China is most likely to achieve technological and organisational upgrading by participating in these networks. From this perspective, in order to effectively participate in the networked global innovation economy, strategic state responses are required in the areas of resource and knowledge management, domestic- and international-level bargaining and alliance construction, and the development of collaborative governance mechanisms. Do you agree with this observation?

8 It has been argued that, via the development of embedded relational ties with key stakeholders, firms can position themselves in the global economy in a way that allows them to gain access to innovative technological information in a bid to gain access to assets that have been intellectually and politically codified (e.g., copyrighted ideas, trademarks, patents and cutting-edge design). This often requires the development of relational assets designed to facilitate the access of firms to the knowledge assets and coordination mechanisms of established lead firms. Does Beijing consider the development of sets of relational assets as important? If so, what strategic relational policy instruments is Beijing developing to help facilitate foreign and domestic technological collaboration?

9 To what extent do you think that Beijing consider pre-competitive market construction, standards development and embedding of Chinese firms in the future organisation of the global innovation and technological architecture important for the future development of the Chinese knowledge economy?

10 Do you think that Beijing has been pressured by international economic actors to conform to a specific model of global economic and technological development that could negatively impact on its future technological development and its position in the global knowledge economy?

11 Could you outline the key policy instruments and governance mechanisms, if any, that Beijing is developing in order to manage/ facilitate the necessary relational, regularity and structural assets and institutional forms required for successful collaborative network development and alliance management in a global context?

12 What role do you think the state has in organising, governing and coordinating such networks? Do you think this role will change in the near future? If so, how?

13 Do you think the overarching result of greater technological and innovative collaboration will be the development of hybrid firms that contains both Chinese and foreign owners?

14 Do you feel that the state loses a degree of control over its domestic firms when they form alliances with foreign firms and become integrated into global production and innovation networks? Does Beijing have any policy directive designed to minimise this or allow for the greater appropriation of technological and financial rents from both domestic and foreign firms operating in China?

15 My book is focusing on the 3G and 4G wireless sector as a case study. This is a sector that is becoming increasingly defined by collaborative

international R&D. Do you think that the movement to 3G and 4G provides the Chinese government with a window of opportunity to be involved in the development of platform architecture, collaborative development and market construction? That is, what role exactly do you envisage for Beijing and domestic Chinese firms in the development of innovation-based wireless technology such as 3G and 4G wireless technology?

16 Recent cooperative linkages via strategic alliances in the Chinese wireless communication sector suggest that China is moving beyond its policy of building national domestic champions to one of global cooperation and competition. Instead of being confined to the government bureaucracy, the decision-making process has become more receptive to the lobbying of business leaders, sectoral and regional interests, and to some extent transnational business groups. Do you agree with this observation?

17 Do you feel this is a sector-specific occurrence? That is, do different sectors require different global management strategies and levels of interdependence and global embeddedness?

18 It has been suggested that a multilevel bargaining and lobbying apparatus is arising at a global level that is comprised of a complex network of supra-national regulatory institutions and a network of multi-stakeholders that are both domestic and international in nature. Do you agree with the assertion?

19 How have Beijing's political and economic bargaining strategies been impacted or modified in response?

20 Do you think that Beijing is developing new transnational political, diplomatic and economic development strategies and regulatory institutions in response?

21 Do you think that a global governance gap is emerging as a direct result of the emerging structure of the global knowledge/innovation economy?

22 Do you consider that the increasing need for alliances and network positioning in the global economy requires the institutional and political modification of the current international trading system?

Appendix D – Brief interviewee descriptions

Feng Xingyuan is a professor of Rural Development at the Chinese Academy of Social Sciences (CASS) in Beijing.
Ge Wu is the Executive Director of China Society of Information Economy, Institute of National Economy, Shanghai Academy of Social Science.

Liu Xielin is a professor in Innovation Management at the School of Management at the University of Chinese Academy of Sciences in Beijing.

Ran Bao is the editor of Digital Media magazine and one of the founders of Beijing Internet TV station.

Sheng Quhong is an associate professor in Innovation and Knowledge Management at Tsinghua University in Beijing.

Wang Hui is an associate professor at the Chinese Academy of Social Sciences (CASS) in Shanghai.

Xiang Ligang is the Vice President and Secretary General of the 3G Industry Association.

Zang Caiming is a PHD graduate at the Center for Informatisation Study at the Chinese Academy of Social Sciences (CASS) in Shanghai.

Zang Maria is an associate professor at the School of Economics and Management (SEM) at Tsinghua University.

Bibliography

Abramowitz, M. (1986). 'Catching up, forging ahead and falling behind', *Journal of Economic History*, 46(2): 385–406.

Accenture, (2012). *Annual Report* 2012. Accenture.

Addendum to FOW Center Report. (2007). From Workshop of the World to Global R&D Center? The Rise of Innovation in China. Addendum to FOW Center Report.

Adner, R. (2012). Avoiding the Blind Spot: How to Multiply Your Odds of Successful Innovation. Harvard Business Publishing.

Adner, R. and Kapoor, R. (2006). Innovation Ecosystems and Innovator Outcomes: Evidence from the Semiconductor Lithography Equipment Industry, 1962–2004. INSEAD Working Paper.

Adner, R. and Kapoor, R. (2010). 'Value creation in innovation ecosystems: how the structure of technological interdependence affects firm performance in new technology generations', *Strategic Management Journal*, 31: 306–33.

Ahrens, N. (2013). China's Competitiveness: Myth, Reality, and Lessons from the United States and Japan: Case Study: Huawei. Center for Strategic and International Studies.

Aiyar, S., Duval, R., Puy, D., Wu, Y., and Zang, L. (2013). Growth Slowdowns and the Middle-Income Trap. International Monetary Fund.

Alcatel-Lucent. (2010). Alcatel-Lucent Annual Report 2010. [Online], Available: http://annual-report.alcatel-lucent.com

Alcatel-Lucent. (2012). Alcatel-Lucent and China Mobile Accelerate Development of LightRadio™ to Support Exploding Customer Demand for Mobile Broadband in China. Alcatel-Lucent, [Online], Available: http://www3.alcatel-lucent.com/wps/portal/newsreleases/detail?LMSG_CABINET=Docs_and_Resource_Ctr&LMSG_CONTENT_FILE=News_Releases_2012/News_Article_002595.xml

Aligica, P.D. and Tarko, V. (2012). 'State capitalism and the rent-seeking conjecture', *Constitutional Political Economy*, 23(4): 357.

Anderson, M. and Gunnarsson, G. (2003). Development and Structural Change in the Asia-Pacific: Globalising Miracles or the End of a Model. Routledge.

Andrews, E.L. and Greenspan, A. (2008). 'Greenspan concedes error in regulation', *New York Times*, 23 October.

Antonelli, C. (2010). 'From population thinking to organisation thinking: coalitions for innovation. A review article of complexity perspectives in innovation and social change', in Lane, D.A. and van Der Leeuw, S.E., Pumain, D. and West, G. (eds), Innovation and Social Change. Springer.

Appelbaum, R.P., Parker, R. and Cao, C. (2008). China's (Not So Hidden) Developmental State: Becoming a Leading Nanotechnology Innovator in the 21st Century. National Science Foundation Grant.

Arbache, J. (2013). In the Special Times of State Capitalism. Special to the BRICS Post. [Online], Available: http://thebricspost.com/in-the-times-of-state-capitalism/#.Uy0KIvmSxsE

Ahrens, N. (2010). Innovation and the Visible Hand: China, Indigenous Innovation, and the Role of Government Procurement. Carnegie Endowment for International Peace.

Asplund, L. and Nanaden, P. (2009). Long Term Evolution. Capgemini: Consulting, Technology Outsourcing.

Atkinson, R. (2012). Enough Is Enough: Confronting Chinese Innovation Mercantilism. The Information Technology and Innovation Foundation.

Ayvazian, B. (2011). LTE-TDD Operator Business Case and Adoption Forecast: White Paper: Heavy Reading, [Online], Available: http://downloads.light-reading.com/wplib/heavyreading/LTETDD_WP_Phase2_final_v3.pdf

Bach, D., Abraham, L., Newman and Weber, S. (2006). 'The international implications of China's fledgling regulatory state: from product maker to rule maker', *New Political Economy*, 11(4): 499–518.

Barber, B. (1995). Jihad vs McWorld: How Globalisation and Tribalism Are Reshaping the World. Ballantine Books.

Barnard and Chaminade. (2012). 'From global production networks to global innovation networks', in Perdeson, T., Bals, L., Orberg, P.D. and Larsen, M.M. (eds), The Off-shoring Strategy: Strategic Design and Innovation for Tomorrow's Organisation. Springer.

Barry, A. (2006). 'Technological zones', *European Journal of Social Theory*, 9(2): 239–53.

Bauer, J.M., Lang, A. and Schneider, V. (2012). 'Innovation policy and high-tech development: an introduction', in Bauer, J.M. et al. (eds), Innovation Policy and Governance in High-Tech Industries. Springer-Verlag Berlin Heidelberg.

Beaver, P. (2011). TD-LTE Explained. *EE Times-Asia*. [Online], Available: http://www.eetasia.com/STATIC/PDF/201110/EEOL_2011OCT11_NET_RFD_TA_01.pdf?SOURCES=DOWNLOAD

Beeson, M. (2004). 'The rise and fall (?) of the developmental state: the vicissitudes and implications of East Asian interventionism', in Low, L. (ed.), Developmental States: Relevancy, Redundancy or Reconfiguration. Nova Science Publishers.

Beeson, M. and Bell, S. (2009). 'The G-20 and international economic governance: hegemony, collectivism or both?', *Global Governance*, 15: 67–86.

Bell, S. (2011). 'Do we really need a new "constructivist institutionalism" to explain institutional change?', *British Journal of Political Science*, 41(4): 883–906.

Bell, S. and Feng, H. (2007). 'Made in China: IT infrastructure policy and the politics of trade opening in post-WTO China', *Review of International Political Economy*, 14(1). 49–76.

Bell, Stephen and Hui Feng (2013). *The Rise of the People's Bank of China: The Politics of Institutional Change*. Harvard University Press.

Bell, S. and Hindmoor, A. (2014). Masters of the Universe but Slave of the Market: Bankers and the Great Financial Meltdown, forthcoming.

Blackman, J. (2010). 'Huawei: enter the dragon', *Mobile News: The UK's Leading Mobile Industry News*. [Online], Available: http://www.mobilenewscwp.co.uk/2010/12/huawei-enter-the-dragon

Block, F. (2008). 'Swimming against the Current: the rise of a hidden developmental state in the United States', *Politics and Society*, 36(2): 169–206.

Boehler, P. (2014). 'China spending more than Europe on science and technology as GDP percentage, new figures reveal', *Southern China Morning Post*. [Online], Available: http://www.scmp.com/news/china-insider/article/1410178/china-spending-more-europe-science-and-technology-gdp-percentage

Boeing, (2012). Commercial Aircraft Corp. of China and Boeing Sign Collaboration Agreement to Partner in Areas Advancing Commercial Aviation Industry Growth. [Online], Available: http://boeing.mediaroom.com/index.php?s=20295&item=2156

Bolesta, A.J. (2012). China as a Post-Socialist Developmental State: Explaining Chinese Development Trajectory. The London School of Economics and Political Science.

Booz and Company. (2012). Booz and Company World Telecommunications Outlook 2012. Booz and Company.

Border Controller Industry News. (2012). 'Alcatel-Lucent CEO welcomes Chinese investment in digital industry,' *Border Controller Industry News*. [Online], Available: http://session-border-controller.tmcnet.com/news/2012/06/25/6394174.htm.

Boyd, R. and Ngo, T.-W. (2006). State Making in Asia. Routledge.

Boyer, R. and Drache, D. (1996). The Limits of Globalisation. Routledge.

Braunstein, E. and Epstein, G. (2002). Bargaining Power and Foreign Direct Investment in China: Can 1.3 Billion Consumers Tame the Multinationals? University of Massachusetts.

Bremmer, I. (2009). 'State capitalism comes of age: the end of the free market', *Foreign Affairs*, 88(3): 40–55.

Bremmer, I. (2010). The End of the Free Market: Who Wins the War between Corporations and States. Portfolio, Penguin Group.

Bremmer, I. (2012). 'The primary purpose of state capitalism is not to produce wealth but to ensure that wealth creation does not threaten the ruling elite's political power', *The Economist's Debates, The Economist*.

Bresson, S. (2004). Capitalism with Chinese Characteristics. The Public, the Private and the International. Working Paper 104. Murdoch Centre.

Breznitz, D. (2012). 'Testimony', US-China Economic and Security Review Commission, 10 May. [Online], Available: http://www.uscc.gov/sites/default/files/5.10.12breznitz.pdf

Brinkerhoff, D. (2005). Organisational Legitimacy, Capacity and Capacity Development. European Centre for Development Policy Management (ECDM) Discussion Paper 58A.

Brismark, G. and Alfalahi, K. (2008). 'Patent strategies–a fork in the road toward 4G', Open Standards How to Work with Intellectual Property, EBR #3.

Bunnell, T.G. and Coe, N.M. (2001). 'Spaces and scales of innovation progress', *Human Geography*, 25(4): 569–89.

Bushell-Embling, D. (2011). China Mobile in TD-LTE Testing R&D Pact, telecomasia.net. [Online], Available: http://www.telecomasia.net/content/china-mobile-td-lte-testing-rd-pact.

Business Monitor. (2011). China Telecommunications Report: Includes 5-Year Forecasts to 2015. Business Monitor International.

Bute, R. (2013). Innovation: The New Mantra for Science and Technology Policies in India, Pakistan and China. Institute of Defence Study and Analysis.

C114. (2009a). 'MIIT on Huawei, ZTE R&D, domestic telecom fees', C114, *Online Media Blog*. [Online], Available: http://www.cn-c114.net/575/a425388.html

C114. (2009b). 'ZTE, Lenovo, Huawei win China Mobile handset contracts', *C114, Online Media Blog*. [Online], Available: http://www.cn-c114.net/575/a425388.html

C114. (2009c). 'China Mobile kicks off bid for TD-LTE terminal R&D', C114, *Online Media Blog*. [Online], Available: http://www.cn-c114.net/583/a432289.html

C114. (2010). 'China sees rosy TD-LTE tech future', C114, *Online Media Blog*. [Online], Available: http://www.cn-c114.net/576/a514804.htmlCll4. (2011). '4G networks spreading faster than expected', *C114, Online Media Blog*. [Online], Available: http://www.cn-c114.net/2503/a583657.html

Cll4. (2011). '4G networks spreading faster than expected', *C114, Online Media Blog*, [Online], Available: http://www.cn-c114.net/2503/a583657.html [26/2/11].

Cai, J. (2011). 'Mobile communications in China: levels of technological dynamism', *Technology Analysis & Strategic Management*, 23(2): 123–43.

Cajian. (2009). 'Grand Plan for 4G and R &D Rollout', *Caijing*, January.

Cao, S. (2011). 'China Mobile ambitious to lead 4G mobile technology', *China Daily*. 11 July. [Online], Available: http://www.chinadaily.com.cn/business/2011–07/11/content_12873730.htm

Cao, C., Suttmeier, R.P. and Simon, D.F. (2006). 'China's 15-year science and technology plan', *Physics Today*, December.

Carayannis, E.G. and Alexander, J. (2006). Global and Local Knowledge: Glocal Transatlantic Public-Private Partnerships for Research and Technological Development. Palgrave Macmillan.

Carney, R.W. (2013). The Stabilising State: State Capitalism as a Response to Financial Globalisation. Japan Studies Program Seminar Series.

CASS. (2011). Chinese Academy of Sciences Has Big Plans for Nation's Research. Institute of Earth Environment. Chinese Academy of Social Sciences.

Castells, M. (1992). 'Four asian tigers with a dragon head: a comparative analysis of the state, economy, and society in the asian pacific rim', in Appelbaum, R. and Henderson, J. (eds), States and Development in the Asia Pacific Rim. Sage.

Cavazos Cepeda, R., Lippoldt, D. and Senft, J. (2010). 'Policy Complements to the Strengthening of IPRS in Developing Countries', OECD Trade Policy Papers, No. 104, OECD Publishing. [Online], Available: http://dx.doi.org/10.1787/5km7fmwz85d4-en.

Chandler, A.D. (199). Scale and Scope: They Dynamics of Industrial Capitalism. The Belknap Press of Harvard University.

Chang, H.-J. (2006). Kicking Away the Ladder: Development in Historical Perspective. Anthem.

Chao, L. (2010). 'Beijing revises purchase policies', *Wall Street Journal*, April.

Cheung, Yin-Wong; Qian, Xingwang, (2009). *The Empirics of China's Outward Direct Investment*, CESifo working paper, No. 2621.

Chen, Y. (2013). 'China shows off scientific, technological achievements'. *China Daily*. September.

Chen, S.-H. (2004). 'Knowledge intensification in Taiwan's IT sector', in Chen, T.J. and Lee, J. (eds), The New Knowledge Economy of Taiwan. Edward Elgar.

Chen, S. (2011). 'China communication technology annual meeting', *TD Forum Weekly News*. [Online], Available: http://www.tdscdma-forum.org/en/weekly/108/.

Chen, Y.L. (2012). 'China to Hold 2-Year Trial for Private Mobile Service Providers', *Bloomberg*, Online Blog. [Online], Available: http://mobile.bloomberg.com/news/2013–01–08/china-to-hold-2-year-trial-for-private-mobile-service-providers.html.

Chen, T.J. and Lee, J. (2004). The New Knowledge Economy of Taiwan, (eds). Edward Elgar.

Chen, W.R., Giemno, J. and Verbeek, A.D. (2012). 'Nokia and the New Mobile Ecosystem: Competing in the Age of the Internet', *Instead*, The Business School of the World.

Cherry, B. (2007). 'The telecommunications economy and regulation as co-evolving complex adaptive systems: implications for federalism', *Federal Communications Law Journal*, 59(2): 369.

Chesbrough, H. (2012). Open Innovation Where We've Been and Where We're Going. Industrial Research Institute.

Chesbrough, H.,Vanhaverbeke, W., Bakici, T. and Lopez, H. (2011). Open Innovation and Public Policy in Europe. Science Business Publishing Ltd.

China Daily. (2011). 'China Mobile ambitious to lead 4G mobile technology', *China Daily*, July. [Online], Available: http://www.chinadaily.com.cn/business/2011–07/11/content_12873730.htm.

China Economic Review. (2011). 'TD-LTE-Advanced tries for 4G', *China Economic Review*, 10 January. [Online], Available: http://www.chinaeconomicreview.com/node/44665.

China Internet Network Information Center. (2013). Statistical Report on Internet Development in China. China Internet Network Information Center.

China Mobile. (2010). China Mobile Limited Annual Report 2010. China Mobile.

China Mobile. (2011a). TDE Industry Briefing: May 2011/9. China Mobile Research Institute.

China Mobile. (2011b). TDE Industry Briefing: May 2011/10. China Mobile Research Institute.

China People's Daily. (2011). 'China in race to launch 4G', *China People's Daily*. [Online], Available: http://english.peopledaily.com.cn/90001/90778/7275044.html.

China.org.com. (2011). China Mobile Ambitious to Lead 4G Mobile Technology. Xinhua, 11 July.

Chinascope Financial. (2013). 'Chinese telecom equipment vendors win bulk of China Mobile's 4G tender', *Chinascope Financial*. [Online], Available: http://www.chinascopefinancial.com/en/news/post/28062.html.

China Tech News. (2009). 'China Mobile plans to build research center in Silicon Valley', *China Tech News*. [Online], Available: http://www.chinatechnews.com/2008/10/10/7707-china-mobile-plans-to-build-research-center-in-silicon-valley.

China Times. (2011). 'China Mobile builds alliances to compete in 4G technology', *China Times*. [Online], Available: http://www.thechinatimes.com/online/2011/07/338.html.

Chinese Ministry of Commerce, (2012). *Chinese OFDI*. Chinese Ministry of Commerce.

Chung, W. and Hamilton, G. (2009). 'Getting rich and staying connected: the organisational medium of Chinese capitalists', *Journal of Contemporary China*, 18(58): 47–67.

Coates, D. (2000). Model of Capitalism: Growth and Stagnation in the Modern Era. Polity Press.

Coenen, L., Suurs, R. and van Sandick, S. (2010). Upscaling Emerging Niche Technologies in Sustainable Energy: an International Comparison of Policy

Approaches. Centre for Innovation, Research and Competence in the Learning Economy (CIRCLE). Lund University.

Cohn, W.A. (2009). 'How free are free markets', *New Presence,* 12(1): 23–29.

Commission on the Theft of American Intellectual Property. (2013). The IP Commission Report. The National Bureau of Asian Research.

Communiqué de presse. (2012). 'Alcatel-Lucent wins a major role in the deployment of China's LT-TDE trial network', *Communiqué de presse.* [Online], Available: http://www.edubourse.com/finance/actualites.php?actu=78775.

Cox, R. (1987). Production Power and World Order: Social Forces in the Making of History. Columbia University Press.

CPC Central Committee and the State Council. (2010). Several Opinions on Further Improving the Work of Utilizing Foreign Investment, 9, April.

CPC Central Committee and the State Council. (2012). The Opinions on Deepening the Reform of Science and Technology Systems and Speed Up the Construction of a National Innovation System. September.

Dalton, M. (2011). 'EU finds China gives aid to Huawei, ZTE', *Wall Street Journal,* February. [Online], Available: http://online.wsj.com/article/SB1000142405274 8703960804576120012288591074.html.

Danneels, E. (2004). 'Disruptive technology reconsidered: a critique and research agenda', *Journal of Product Innovation Management,* 21: 246–58.

Darner, E. and Pettit, J. (2012). *The Myth of First Mover Advantage.* IHS Consulting.

Datang Telecom Technology & Industry Group. (2010a). Datang Annual Report 2010. China Datang Mobile Corporation.

Datang Telecom Technology & Industry Group. (2010b). Datang Launched 300 Patent Applications in 1st half of 2010 Expediting Internationalisation of TD Industry. China Datang Mobile Corporation.

Datang Telecom Technology & Industry Group. (2011). About Datang. China Datang Mobile Corporation.

Davenport, T., Leibold, M. and Voelpel, S. (2006). Strategic Management in the Innovation Economy: Strategy Approaches and Tools for Building Innovation Capabilities. Wiley.

Deans, P. (2004). 'The People's Republic of China: the post-socialist developmental states', in Low, L. (ed.), Developmental States, Redundancy or Reconfiguration. Nova Publishers.

DeGrasse, M. (2012). 'LTE Patent Pool Launches with 10 Companies', *RCR Wireless US.* Blog. [Online], Available: http://www.rcrwireless.com/article/20121003/ featured/lte-patent-pool-launches-10-companies/.

Dent, C.M. (2004). The New International Political Economy of East Asia and the Developmental State, in Low, L. (ed.), Developmental States, Redundancy or Reconfiguration. Nova Publishers.

Depret, A. and Hamdouch, M.H. (2011). 'Sustainable development and the territorial dynamics of the "green economy", actors, scales and policies', *Portuguese Journal of Geography.*

Deyo, F. C. (1987). The Political Economy of the New Asian Industrialism. Cornell.

Dicken, P., Kelly, P.,Olds, K. and Yeung, H.W. (2001). 'Chains and networks, territories and scales: towards a relational framework for analysing the Global Economy', *Global Networks,* 1(2): 89–112.

Dosi, G, (1982). 'Paradigms and technological trajectories. A suggested interpretation of the determinants and directions of technical change', *Research Policy*, 11(3): 47–162.

Doyletech Corporation and D.R. Senik and Associates Inc. (2007). Wireless Technology Roadmap: 2006–2016. Information and Communications Technology Council (ICTC).

Dunning, J.H. (1997a). Alliance Capitalism and Global Business. London and New York: Routledge.

Dunning, J.H. (1997b). Alliance Capitalism in International Business and the World Economy. Routledge.

Eberhardt, M., Helmers, C. and Zhihong, Y. (2011). Is the Dragon Learning to Fly? An Analysis of the Chinese Patent Explosion. The University of Nottingham Research Paper No. 2011/16.

Economic Intelligence Unit, (2011). *Multinational companies and China: What future*. The Economist.

Eisenhardt, K. and Santos, F.M. (2001). 'Knowledge-based view: a new theory of strategy?', in Pettigrew, A., Thomas, H. and Whittington, R. (eds), Handbook of Strategy and Management. Sage Publications.

EMT Worldwide. (2011). Accelerating Development of Next-generation Wireless Technology in China. IML Group. [Online], Available: http://www.epdtonthenet.net/article/45668/Accelerating-development-of-next-generation-wireless-technology-in-China.aspx.

Enright, M.J., Scott, E.E. and Chang. K. (2005). The Greater Pearl River Delta and the Rise of China. John Wiley.

Ernst, D. (2002). 'Global production networks and the changing geography of innovation systems. implications for developing countries', *Economics of Innovation and New Technology*, 11(6): 497–523.

Ernst, H. (2003). 'Patent information for strategic technology management', *World Patent Information,* 25: 233–42.

Ernst, D. (2008). 'Can Chinese IT firms develop innovative capacities within global knowledge networks?', in Rowen, H., Gong, M., Hancock and Miller, W.F. (eds), Greater China's Quest for Innovation. The Walter H. Shorenstein Asia-Pacific Research Center, 197–216.

Ernst, D. (2010). Indigenous Innovation and Globalisation–the Challenge for China's Standardization Strategy. East-West Center, Honolulu.

Ernst, D. and Naughton, B. (2008). 'China's emerging industrial economy: insights from the IT industry', in McNally, C.A. (ed.), China's Emergent Political Economy: Capitalism in the Dragon's Lair. Routledge.

Escai, H. (2009). Trade Collapse, Trade Relapse and Global Production Networks: Supply Chains in the Great Recession. OECD Roundtable on Impacts of the Economic Crisis on Globalisation and Global Value Chains. Conference Paper.

ETSI IPR Online Database, (2010). LTE Essential Patent Owners 2010. [Online], Available: http://www.etsi.org/WebSite/AboutETSI/LegalAspects/iprdb.aspx.

Evans, P. (1995). Embedded Autonomy: States and Industrial Transformation. Princeton University Press.

Evans, A. (2003). 'Taming the counterfeit dragon: the WTO, trips and Chinese amendments to intellectual property', *Laws*, 31: 587.

Fan, P. (2006). 'Catching up through developing innovation capability: evidence from China's telecom-equipment industry', *Technovation,* 26: 359–68.

Fan, P. (2010). 'Catching-up through staged development and innovation: The case of Chinese telecom companies', *Journal of Science and Technology Policy in China*, 1(1): 64–91.

Faulhaber, G. and Farber, D.J. (2010). 'Innovation in the wireless ecosystem: a customer-centric framework', *International Journal of Communication*, 4: 73–112.

Feng, X. (2011). Professor, Rural Development Institute Chinese Academy of Social Science. Personal Interview, Beijing China. 15 April.

Fensterheim, D., Huang, Y. and Murray, F. (2009). Republic of Government Vis-À-Vis Republic of Science: Analyzing China's Scientific Knowledge Production, CDDRL Working Papers, 27 April.

Fewsmith, J. (2007). 'China under Hui Jinto', *China Leadership Monitor*, 14.

Fomin, V.V., Su, J. and Gao, P. (2011). 'Indigenous standard development in the presence of dominant international standards: the case of the AVS standard in China', *Technology Analysis & Strategic Management*, 23(7): 745–58.

Freeland, C. (2010). 'China's Economic Model isn't the Answer for the U.S.', *Washington Post*, 10 August. [Online], Available: http://www.washingtonpost.com/wp-dyn/content/article/2010/08/29/AR2010082902898.html.

Frost and Sullivan. (2011). 'Frost & Sullivan Recognises ZTE as 2011 LTE Vendor of the Year', C114, Online Media Blog. [Online], Available: file:///users/victoriahiggins/Desktop/PHD/Frost%20&%20Sullivan%20Recognises%20ZTE%20as%202011%20LTE%20Vendor%20of%20the%20Year%20-%204G%5Ccn-c114%20¡ª%20C114%20-%20China%20Communication%20Network.html.

Gabriele, A. and Haider Khan, A. (2008). Enhancing technological progress in a market-socialist context: China's national innovation system at the crossroads. Unctad – United Nations, JKSIS.

Gaberil, C. (2010). 'Huawei leads field for first stage LTE contracts', *Rethink Wireless*, 18 December. [Online], Available: http://www.rethink-wireless.com/2010/11/25/huawei-leads-field-stage-lte-contracts.htm.

Gaberil, C. (2011). 'Huawei and ZTE support 3's 4G push', *Rethink Wireless*, 26 January. [Online], Available: http://www.rethink-wireless.com/2011/1/26/Huawei-and-ZTE-support-3s-4G-push.htm.

Gabi, D. and Keating, C. (1996). Global Communication Alliances, Forms and Characteristics of Emerging Organisations. International Bureau.

Gallagher, K.S. (2003). Foreign Technology in China's Automobile Industry: Implications for Energy, Economic Development, and Environment. China Environment Series. Woodrow Wilson Center for International Scholars, Washington, DC.

Gao, X. and Liu, J. (2012). 'Catching up through the development of technology standard: the case of TD-SCDMA in China', *Telecommunications Policy*, 36: 531–45.

Gay, B. (2008). Firm Dynamic Governance of Global Innovation by Means of Flexible Networks of Connections. MPRA Paper No. 12525. [Online], Available: http://mpra.ub.uni-muenchen.de/12525/1/MPRA_paper_12525.pdf.

Ge, W. (2011). Executive Director of China Society of Information Economy, Institute of National Economy, Shanghai Academy of Social Science. Personal Interview. Shanghai, April 2011.

Geels, F.W. (2004). 'From sectoral systems of innovation to socio-technical systems: insights about dynamics and change from sociology and institutional theory', *Research Policy*, 33: 897–920.

Geels, F.W. and Schot, J. (2007). 'Typology of socio-technical transition pathways', *Research Policy*, 36: 399–417.

Gereffi, G. and Wyman, D. eds., (1990). *Manufacturing Miracles*. Princeton University Press.

Gereffi, G., Humphrey, J., Kaplinsky, R. and Sturgeon T.J. (2001), 'Introduction'', in Gereffi, G. et al. (eds), Globalisation, Value Chains and Development. IDS Bulletin, *Institute of Development Studies, Sussex*, 32(2): 1–8.

Gerschenkron, A. (1962). *Developing Nations in Historical Perspective*. Belknap Press of Harvard.

Gilboy, G.J. (2004). 'The myth behind China's miracle', *Foreign Affairs*, July/August, 33–48.

Gill, S. (1995). 'Globalisation, market civilization and disciplinary neo-liberalsim', *Millennium: Journal of International Studies*, 34: 399–423.

Gill and Kharas (2007Gilpin, R. (2001). Global Political Economy: Understanding the International Economic Order. Princeton University Press.

Global Telecom Business. (2010). Interview: Shilirong of ZTE. [Online], Available, http://search.proquest.com/docview/756205797.

Godinhol, M.M. and Ferreira, V. (2013). Two Emerging Innovative Dragons: An Analysis of the IPR Strategy of China's Huawei and ZTE. Proceedings of PICMET'13 Technology Management for Emerging Technologies.

Goedeking, P. (2010). Networks in Aviation Strategies and Structures: Strategies and Structures. Springer.

Goldin, I. and Reinert, K.A. (2007). Globalisation for Development: Trade, Finance, Aid, Migration, and Policy. World Bank Publications.

Goldman-Sachs Global Investment Research. (2011). TD-LTE: Gearing up to Cover 2.7bn People in Asia by 2013. Equity Research. Goldman Sachs.

Gonzalez-Vicente, R. (2011). 'The internationalisation of the Chinese state', *Political Geography*, 30: 402–11.

Gou, Z. (2006). IT innovation and the development of information industry', *Management Forum*. [Online], Available: http://unpan1.un.org/intradoc/groups/public/documents/APCITY/UNPAN027039.pdf.

Graham, I. (2008). China EU Information Technology Standards Research Partnership. European Commission within the Seventh Framework Programme (2007–2013).

Granovetter, M. (1985). 'Economic action and social structure: the problem of embeddedness', *American Journal of Sociology*, 1(3): 481–510.

Gregory, N.F., Nollen, S.D. and Tenev, S. (2009). New Industries from New Places: The Emergence of the Software and Hardware Industries in China and India. Stanford University and the World Bank.

Grinnan, T. (2011). *China Telecom Equipment Policy Shift Drives Upgrade*. HSBC Global Research. Hong Kong and Shanghai.

Grivolas, J. (2011). Global Opportunities for LTE TDD. Ovum.

Guangzhou LiLon Consulting & Service Co., Ltd. (2011). China Patent News in the First Half Year of 2011. Guanzhou LiLon Consulting & Services Co. Ltd.

Guo, Z. (2006). IT innovation and the development of information industry', *Management Forum*, [Online], Available: http://www.china-cic.org.cn/english/digital%20library/200602/1.pdf

Haberly, D. (2011). Strategic Sovereign Wealth Fund Investment and the New Alliance Capitalism: A Network Mapping Investigation. Unpublished manuscript.

Hagel, J. (2008). 'Creation Nets: Harnessing the potential Of Open Innovation', *Journal of Service Science*, 1(2): 27–40

Hagel, J. and Brown, J.S. (2005). The Only Sustainable Edge. Harvard Business School Press.

Hagel, J., Brown, J.S. and Jelinek, M. (2010). Relational Networks, Strategic Advantage: New Challenges for Collaborative Control. [Online], Available: http://www.johnseelybrown.com/RelationalNet.pdf.

Hagerdooorn, J. (2003). 'Sharing intellectual property rights – an exploratory study of joint patenting amongst companies', *Industrial and Corporate Change*, 12(5): 1035–50.

Hall, B.H. (2010). 'Open innovation & intellectual property rights: the two-edged sword', *Japan Spotlight*, Jan/Feb, 18–19.

Hallikas, J. (2005) cited in Yu, L., Suojapelt, K., Hallikas, J. and Tang, O. (2008) (eds), 'Chinese ICT industry from supply chain perspective: A case study of the major Chinese ICT players', *International Journal of Production Economics*, 15: 374–87.

Han, K., Oh, W., Shin, I.K., Chang, R.M., Oh, H. and Pinsonneault, A. (2012). Value Co-Creation and Wealth Spillover in Open Innovation Alliances. School of Business, Yonsei University.

Hannan, M.T., Carroll, G.R. and Polos, L. (2003). 'The organisational niche', *Sociological Theory*, 21(4): 309–40.

Harwit, E. (2007). 'Building China's Telecommunications Network: Industrial Policy and the Role of Chinese State-Owned, Foreign and Private Domestic Enterprises,' *The China Quarterly*, 190: 311–332.

Hausknost, D. and Haas, W. (2013). The Role of Innovation in a Socio-ecological Transition of the European Union. Neujobs Working Paper NO. D 1.4

Hays, D. (2011) cited in Hagel, J., Brown, J.S. and Jelinek, M. (2010) (eds), Relational Networks, Strategic Advantage: New Challenges for Collaborative Control. [Online], Available: http://www.johnseelybrown.com/RelationalNet.pdf.

Hays, D. (2011). 'Reality check: wave of consolidation washing over the wireless'. *RCR Wireless U.S. Intelligence on All Things Wireless*. [Online], Available: http://www.rcrwireless.com/article/20110104/reality_check/reality-check-wave-of-consolidation-washing-over-wireless/.

Heires, M. (2008). 'The International Organisation for Standardisation (ISO)', *New Political Economy*, 13:3.

Hekkerta, M.P., Suursa, R.A.A., Negroa, S., Kuhlmanna, B.R.E.H.M. and Smitsa, A. (2006). Functions of Innovation Systems: A New Approach for Analysing Technological Change. Utrecht University, Copernicus Institute for Sustainable Development and Innovation, Department of Innovation Studies.

Henderson, R.M., and Clark, K. B. (1990). Architectural innovation: The reconfiguration of existing product technologies and the failure of established firms'', *Administrative Science Quarterly*, 35(1): 9–22.

Henkel, J.,Baldwin, C.Y. and Shih, W.C. (2012). 'IP modularity: profiting from innovation by aligning product architecture with intellectual property', Working Paper, 13–012. Harvard Business School Working Papers.

Hire, S. (2009). 'From TD-SCDMA to TD-SCDMA-LTE', *Presentation to 4G Wireless Broadband Evolution Seminar.* Hong Kong, 7 September.

Hire, S. (2011) cited in Quing, L.A. 'China's TD-LTE spreads across globe', *ZDNet Asia*, 22 June 2011. [Online], Available: http://www.zdnetasia.com/chinas-td-lte-spreads-across-globe-62300869.htm.

Hobday, M. (1998). Innovation in East Asia: The Challenge to Japan. Cheltenham: Edward Elgar.

Hongi, Y. (2011). 'Reading the Twelfth Five-Year Plan: China's communication-driven mode of economic restructuring', *Journal of Communication*, 5: 1045–57.

Hou, J. (2011). The Role of MNC's in China's Standardisation. Economic Series. No. 114. East-West Center Papers.

Hsueh, R. (2011). China's Regulatory State: A New Strategy for Globalisation. Ithaca, NY: Cornell University Press.

Hu, A.G. (2010). 'Propensity to patent, competition and china's foreign patenting surge', *Research Policy*, 39: 985–993.

Hu, M.C. and Mathews, J.A. (2008). 'China's national innovative capacity', *Research Policy*, 37(9): 1465–79.

Huang, B. (2011a). 'The great leap forward: How the world's largest operator aims to jump one generation', China Mobile. Interview by Thorén, M. and Harvard, A. (2011). EBR #2.

Huang, X. (2011b). 'China Mobile builds alliances to compete in 4G technology', *China Times*. [Online], Available: http://www.thechinatimes.com/online/2011/07/338.html.

Huang, X. (2011c). 'China Mobile ambitious to lead 4G tech China Daily', China Mobile's Research Institute, *China.org.com*, July. [Online], Available: http://www.china.org.cn/business/2011–07/11/content_22961480.htm.

Huawei. (2010a). Huawei Annual Report 2010. Huawei Technologies Co., Ltd.

Huawei. (2010b). The LTE Dream is Coming True. Huawei Technologies Co., Ltd.

Huawei, (2011). *Company Annual Report 2011*. Huawei.

Huawei, (2012). Huawei Annual Company Report 2012. Huawei.

Huivao, W. (2010). China's National Talent Plan: Key Measures and Objectives. Brookings.

Hwang, A. (2013). 'China February mobile phone users grow to 1.132 billion, says MIIT', *DIGITIMES*, Taipei, 25 March.

Ikenberry, J. (2008). 'The rise of China and the future of the West: can the liberal trading regime survive,' *Foreign Affairs*, January/February.

Irwin Crookes, P.C. (2008). 'China's Drive Towards an Innovation Economy and the Role of Intellectual Property Regulations: Progress and Tensions in the Software and Software Services Markets', *CEA Annual Conference*, University of Cambridge.

Jacobides, M., Knudsen, and Augier, M. (2006). 'Benefiting from innovation: value creation, value appropriation and the role of industry architecture', *Research Policy*, 35: 1200–21.

James, W. (1912). Essays in Radical Empiricism. New York: Longman Green and Co.

Janeway, W. H. (2012). *Doing Capitalism in the Innovation Economy: Markets, Speculation and the State*. Cambridge University Press.

Jayasuriya, K. (2004). 'The new regulatory state and relational capacity', *Policy & Politics*, 32(4): 487–501.

Jessop, B. (1997). 'The governance of complexity and the complexity of governance: preliminary remarks on some problems and limits of economic governance', in Amin, A. and Hau, J. (eds), *Beyond Market and Hierarchy*. Edward Elgar.

Jiabao, W. (2012). Answers Questions at the Opening Ceremony of the 6th Summer Davos Forum and Meeting with Business Representatives. Embassy the Peoples Republic of China.

Jingting, S. (2011). '4G networks spreading faster than expected', *China Daily*. 25 February. [Online], Available: http://english.peopledaily.com. cn/90001/90776/90882/7299774.htmlat.

JLJ Group. (2013). China Research & Development Incentives. JLJ Group.

Johnson, C. (1982). MITI and the Japanese Miracle: The Growth of Industrial Policy, 1925–75. Stanford University Press.

Johnson, C. (1995). Japan Who Governs? The Rise of the Developmental State. Norton.

Joia, J. (2009). Connect + Develop with Procter and Gamble. Innovation management. [Online], Available: http://www.innovationmanagement.se/2009/05/28/connect-develop-with-procter-a-gamble/.

Kai, M. (2006). *The 11th Five-Year Plan: Targets, Paths and Policy Orientation*. National Development and Reform Commission.

Kassner, G. (2012). 'China's IP reform: state interests align with intellectual property protection (again)', *Harvard Journal of Law and Technology*, Jolt Digest. [Online], Available: http://jolt.law.harvard.edu/digest/patent/chinas-ip-reform-state-interests-align-with-intellectual-property-protection-again.

Katz, D. and Kahn, R.L. (1966). The Social Psychology of Organisation. New York: Wiley.

Kenis, P. and Provam, K. G. (2005). 'Modes of network governance: structure, management, and effectiveness', *Journal of Public Administration Research and Theory*, 18: 229–52.

Kennedy, S. (2005). The Business of Lobbying in China. Harvard University Press.

Kennedy, S. (2006). 'The political economy of standards coalitions: explaining China's involvement in high-tech standards wars', *Asian Policy*, 2: 41–62.

Kennedy, S. (2007). 'Transnational political alliances an exploration with evidence from China', *Business & Society*, 46(2): 174–200.

Khanna, P. and Khanna, A. (2010). Foreign Policy: A Predictable Future for Technology. NPR.

Kitson, M. and Michie, J. (1998). 'Markets competition and innovation', in Michie, J. and Greive Smith, J. (eds), *Globalisation, Growth and Governance: Creating an Innovative Economy*. Oxford University Press, chapter 5.

Kozhikode, J.L. (2009). 'Developing new innovation models: shifts in the innovation landscapes in emerging economies and implications for global R&D management', *Journal of International Management*, 15: 328–39.

Kuznar, A. (2012). Intangibles in Economies and International Trade. Warsaw School of Economics.

Kwak, J., Lee, H. and Chung, D.B. (2012). 'The evolution of alliance structure in China's mobile telecommunication industry and implications for international standardisation', *Telecommunications Policy*, 36(10–11): 966–76.

Kwak, J., Lee, H. and Fomin, V.V. (2011). 'The governmental coordination of conflicting interests in standardisation: case studies of indigenous ICT standards in China and South Korea', *Technology Analysis & Strategic Management*, 23(7): 789–805.

Kwok-wah, Y. (2012). The Uniqueness of China's Development Model, 1842–2049. World Scientific Publishing Company.

Labonte, M. (2010). The Size and Role of Government: Economic Issues. Congressional Research Service.

Lall, S. (2003). 'Indicators of the relative importance of IPR's in developing countries', *Research Policy*, Elsevier, 32(9): 1657–80.

Lall, S. (2004). 'Reinventing industrial strategy: the role of government policy in building industrial competitiveness', G-24 Discussion Paper Series 28, United Nations Conference on Trade and Development.

Lall, S. (2007). Reinventing Industrial Strategy: The Role of Government Policy in Building Industrial Competitiveness. The Intergovernmental Group on Monetary Affairs and Development.

Lane, D. and Maxfield, R. (1996). 'Strategy under complexity: fostering generative relationships', *Long Range Planning*, 2(29): 215–31.

Laperrouza, M. (2004). China's Broadband Wireless Industry-A Prospective Approach. London School of Economics and IMD.

Lazonick, W. (1991). Business Organisation and the Myth of the Market Economy. Cambridge University Press.

Lee, M. and Lin, D. (2011). 'China's ZTE Eyes 30% Revenue Growth', *Reuters*. [Online], Available: http://www.reuters.com/article/2010/05/17/us-summit-zte-idUSTRE64G26D20100517.

Lee, H. and Sangjo, O. (2008). 'The political economy of standard setting by Newcomers: China's WAIP and South Korea's WIPI', *Telecommunications Policy*, 32: 662–71.

Lee, C.K. and Saxenian, A. (2007). 'Coevolution and coordination: a systemic analysis of the Taiwanese information technology industry', *Journal of Economic Geography*, 8(2), 157–80.

Lei, Z., Sun, Z. and Wighty, B. (2012). Patent Subsidy and Patenting in China. Department of Energy and Mineral Engineering and the EMS Energy Institute. OECD Paris, 28–29 November

Levy, D. (2007). 'Political contestation in global production networks', *Academy of Management Review*, 33(4): 943–63.

Li, M. (2005). 'The rise of China and the demise of the capitalist world-economy: exploring historical possibilities in the 21st century', *Science & Society*, 69(3): 420–48.Ligang, X. (2011). Vice President, Secretary General, 3G Industry Association. *Personal Interview*, April, 12.

Linden, C. (2004). 'China's standard time', *Business and Politics*, 6(3): 1–25.

Ling, Y. (2006). Industrial Policy, FDI and the Development of Local Technological Capacity: A Comparative Analysis of Two Manufacturing Industries in China. Unpublished paper.

Little, D. (2008). Are Suppliers Poised for a Turnaround. Telecom and Media Viewpoint.

Liu, X. (2013). Professor and Associate Dean of the School of Management, University of Chinese Academy of Sciences (GUCAS). Personal Interview. May.

Li-Wen, Lin and Milhaupt, C.J. (2011). We are the (National) Champions: Understanding the Mechanisms of State Capitalism in China. [Online], Available: http://papers.ssrn.com/sol3/papers.cfm?abstract_id=1952623.

Lo, C. and Lewis, C. (2011). 'China Mobile says total subscribers rose to 638.89 million in October', *Reuters*. [Online], Available: http://www.reuters.com/article/2011/11/21/us-chinamobile-idUSTRE7AK03820111121.

Lovelock, P. and Ure, J. (1998). Telecommunications Policy-making in China: a Two Tier Bargaining Model. Center of Asian Studies, University of Hong Kong.

Low, B. (2010). 'An institutional and network perspective of organisational legitimacy empirical evidence from China's telecommunications market', *Asian Academy of Management Journal*, 15(2): 117–34.

Low, B. and Johnston, W. (2006). Knowledge Networks and Organisational Network Legitimacy: Lessons from China's Emerging TDSCDMA Mobility Technology. IMP Group.

Low, B.W. and Johnston, J. (2010). 'Organisational network legitimacy and its impact on knowledge networks: the case of China's TD-SCDMA mobility technology', *Journal of Business & Industrial Marketing*, 25(6): 468–77.

Low, B., Johnston, W. and Wang, J. (2007). 'Securing network legitimacy in China's telecommunication market', *Journal of Business & Industrial Marketing*, 22(2): 97–106.

McCarthy, D. A. (2012). *State Capitalism and Competitive Neutrality*. U.S.-Asia Business Summit. Washington, DC.

Maistre, R. (2011). 'Austria dumps NSN for ZTE', *Light Reading Europe*. [Online], Available: http://www.lightreading.com/document.asp?doc_id=204559.

Mallinson, K. (2011). 'Analyst angle: no consensus on which on which patents are essential to LTE', *RCR Wireless*. [Online], Available: http://www.rcrwireless.com/article/20111116/opinion/analyst-angle-no-consensus-on-which-patents-are-essential-to-lte/.

Marletto, G. (2011). 'Structure, agency and change in the car regime: a review of the literature,' *European Transport* [*Trasporti Europei*], 47: 71–88.

Mars, M.M., Bronstein, J.L. and Lusch, R.F. (2012). 'The value of a metaphor: organisations and ecosystems', *Organisational Dynamics*, 41(4): 271–280.

Martens, B. (2004). The Cognitive Mechanics of Economic Development and Institutional Change. Routledge.

Maltti, W. and Buthe, T. (2003). 'Setting International Standards: Technological Rationality or Primacy of Power', *World Politics*, 1(56): 1–42

Mazzucato, M. (2011). The Entrepreneurial State. Demos.

McGregor, J. (2010). China's Drive for Indigenous Innovation – A Web of Industrial Policies. Global Intellectual Property Group.

McKenzie, D. and Janeway, W.H. (2011). 'Venture capital funds and the public equity market', *Accounting and Finance*, 51(3): 764–68.

McKinsey & Company. (2013). *Changing Rules of the Road for China's Auto Industry Domestic Car Makers Face a Tough Struggle to Adapt to their Evolving Marketplace.* [Online], Available: http://www.mckinseychina.com/changing-rules-of-the-road-for-chinas-auto-industry.

Min, L. (2011). 'Business opportunities in TD-LTE', *ZTE Technologies*, 13(1), 132: 10–13.

Mishra, P. (2012). 'The one capitalism that dare not speak its name', *Bloomberg*, July. [Online], Available: http://www.bloomberg.com/news/2012-07-22/the-one-capitalism-that-dare-not-speak-its-name-pankaj-mishra.html.

Mitchell, W. (2003). 'Searching for theories of dynamic relationships in business strategy: comment on John Dunning's "relational assets, networks and international business activity" paper', *Economics, Culture, and Human Resources Advances in International Management*, 15: 57–66.

Mitleton-Kelly, E. (2011). 'Identifying the multi-dimensional problem space and co-creating an enabling environment', *Emergence: Complexity & Organisation,* 13(1–2): 3–25.

Mobile Business Briefing. (2011). 'China Mobile and SK telecom sign joint R&D deal', *RCR Wireless.* [Online], Available: http://www.allitwares.com/news_topics2-China-Mobile-and-SK-Telecom-sign-joint-RD-deal-632.html.

Mobile Communications International. (2011). 'The time has come for TD-LTE: LTE: the next stage of evolution', *Mobile Communications International,* 24–25. [Online], Available: http://content.yudu.com/Library/A1ub6f/MobileCommunications/resources/25.htm.

Mobile News. (2010). 'Huawei: enter the dragon', *Mobile News.* [Online], Available: http://www.mobilenewscwp.co.uk/2010/12/20/huawei-enter-the-dragon/.

Moody, A. (2011). 'R&D on the march', *China Daily,* 27 December.

Moody, A. and Yan, Y. (2010). 'Cutting edge of 4G standard', *China Daily European Weekly.* [Online], Available: http://europe.chinadaily.com.cn/china/2010–12/10/content_11682848.htm.

Moore, J.F. (1993). 'Predators and prey: a new ecology of competition', *Harvard Business Review,* May–June.

Moore, T.G. (2002). China in the World Market: Chinese Industry and International Sources of Reform in the Post-Mao Era. Cambridge University Press.

Moore, M. (2004). 'The development of political underdevelopment', in Harrison, G. (ed.), *Global Encounters: International Political Economy, Development and Globalisation,* 21–40. Palgrave Macmillan.

Moore, J.F. (2012). 'Predators and prey: a new ecology of competition', *Harvard Business Review.*

MOST. (2004). Preparation of China's National Medium & Long-Term S&T Development Plan and Its Progress, White Paper by Ministry of Science and Technology.

Motorola. (2010). 'TD-LTE; Exciting Alternative', Global Momentum: White Paper. Motorola.

Muncaster, P. (2012). 'China's broadband population is SHRINKING More get their mobiles out to get online', *The A Register.* [Online], Available: http://www.theregister.co.uk/2012/07/23/ china_broadband_numbers_shrinking/.

Murthy, R. (2011). 'China on patents over-drive', *Online AsiaTimes.* [Online], Available: http://www.atimes.com/atimes/China_Business/MA07Cb01.html.

Musacchio, A. (2012). 'State capitalism'. *The Economist.*

Nakayama, I. (2010). Open Innovation and Intellectual Property. Japan Patent Office Asia-Pacific Industrial Property Center.

Naoi, M. (2008). 'Public vs. private enforcement of trade agreements evidence from Chinese trade', Paper prepared for the Conference on Theories of International Political Economy and China, Beijing, China, 8–12 December.

Narayanan, V.K. and Colarelli O'Conner, G. (2010). Encyclopedia of Technology and Innovation Management. Wiley-Blackwell.

Narulal, R. and Dunning, J.H. (1998). 'Explaining international R&D alliances and the role of governments', *International Business Review,* 4(7): 377–97.

National Telecommunication Metrology Station. (2006). *TD-SCDMA Milestone.* [Online], Available: www.emcite.com.

Naughton, B. (1991). 'Hierarchy and the bargaining economy: government and enterprise in the reform process', in Lampton, D.M. and Lieberthal, K.D.

(eds), Bureaucracy, Politics, and Decision Making in Post-Mao China. Berkeley: University of California Press.

Naughton, B. (1995). Growing Out of the Plan: Chinese Economic Reform, 1978–1993. Cambridge University Press.

Naughton, B. (2005). 'The new common economic program: China's eleventh five year plan and what it means,' *China Leadership Monitor*, Issue 16, Fall.

Naughton, B. and Segal, A. (2003). 'China in search of a workable model: technology development in the new millennium', in Keller, W.W. and Samuels, R.J. (eds), Crisis and Innovation in Asian Technology. Cambridge University Press, 160–86.

Nehausler, P., Frietsch, R., Schubert, T. and Blind, K. (2011). 'Patents and the financial performance of firms – An analysis based on stock market data', Fraunhofer ISI Discussion Papers Innovation Systems and Policy Analysis, No. 28. Karlsruhe.

Nelson, R.R. (2003). 'On the complexities and limits of market organisation', *Review of International Political Economy*, 10(4): 697–710.

NGMN. (2006). The NGMN Alliance at a Glance. NGMN.

Nokia Siemens Networks. (2011). 'LTE: the Next Stage of Evolution', *The Future of Wireless*, 173, October.

Norman, R. and Ramírez, R. (2000). 'From value chain, to value constellation: designing interactive strategy', *Harvard Business Review on Managing the Value Chain, 185–220*. Harvard Business School Press.

O'Brien, J. (2014). Enter the Dragon I: The Red Dawn of State Capitalism, Corporate Power and the Search for Villains. Center for Law, Markets and Regulation. [Online], Available: http://www.clmr.unsw.edu.au/article/accountability/regulatory-design/enter-dragon-i-red-dawn-state-capitalism-corporate-power-and-search-villains.

Odling-Smee, F.J. (1988). 'Niche-constructing phenotypes', in Plotkin, H.C. (ed.), The Role of Behaviour in Evolution. MIT Press, 73–132.

OECD. (2005). Governance of Innovation Systems: Synbook Report. OECD.

OECD. (2006). OECD Investment Policy Reviews: China, 2006. OECD.

OECD. (2007). OECD Reviews of Innovation Policy China: Synbook Report. OECD.

OECD. (2008). 'Open Innovation in Global Networks', in Open Innovation in Global Networks. OECD Publishing. [Online], Available: http://dx.doi.org/10.1787/9789264047693-4-en.

OCED, (2009). OECD Patent Statistics Manual. OCED.

OECD. (2011a). World Intellectual Property Report The Changing Face of Innovation. OECD.

OECD. (2011b). A New OECD Project Sources of Growth: Intangible Assets. OECD.

OECD. (2012). Science, Technology and Industry Outlook. OECD.

Office of Management and Budget. (2009). Budget of the US Government Fiscal year 2010. US Government Printing Office.

O'Riain, S. (2000). 'The flexible developmental state: globalisation, information technology and the Celtic Tiger', *Politics and Society*, 28(2): 157–93.

O' Riain, S. (2004a). The Politics of High Tech Growth: Developmental Network States, the Global Economy (Structural Analysis in the Social Sciences 23). New York/Cambridge: Cambridge University Press.

O'Riain, S. (2004b). 'State, competition and industrial change in Ireland 1991–1999', *Economic and Social Review*, 35(1): 27–54.

O' Riain, S. (2009). 'Addicted to growth: developmental statism and neoliberalism in the Celtic Tiger, in M. Bøss (ed.), The Nation-State in Transformation: The Governance, Growth and Cohesion of Small States under Globalisation. Aarhus University Press.

Ovum, (2012). 'Ovum predicts LTE TDD to go mainstream rising to 25 per cent of all LTE connections by 2016', *Fierce Wireless*. [Online], Available: http://www.fiercewireless.com/press-releases/ovum-predicts-lte-tdd-go-maintstream-rising-25-cent-all-lte-connections-201.

Pagani, M. and Fine, C.H. (2008). 'Value network dynamics in 3G–4G wireless communications: A systems thinking approach to strategic value assessment', *Journal of Business Research'*, 6: 1102–12.

Palmberg, C. and Pajarinen, M. (2008). Alliance Capitalism and the Internationalisation of Finnish Firms. Keskusteluaiheita – Discussion Papers.

Pan, C. (2009). 'What is Chinese about Chinese businesses? Locating the "rise of China" in global production networks', *Journal of Contemporary China*, 18(58): 7–25.

Panano, U. and Rossi, M.A. (2009). 'The crash of the knowledge economy', *Cambridge Journal of Economics*, 33: 665–83.

Paolini, M. (2012). Technology to drive wireless disruption, with service monetisation mired in uncertainty. A survey of mobile operators' view of changes in the wireless industry. Senza Filli Consulting. Radisys.

Patrucco, P. P. (2008). 'Collective knowledge production costs and the dynamics of technological systems', *Economics of Innovation and New Technology*, 17(1).

Patrucco (2010). 'Innovation as Loosely Coupled Networks', Paper Presented at Druid Summer Conference.

Pei, M. (2006). China's Tapped Transition: the Limits of Developmental Autocracy. Harvard University Press.

People's Daily. (2009a). 'China's telecom sector gets 3G licenses', *People's Daily*, January. [Online], Available: http://english.peopledaily.com.cn/90001/90776/90881/6569828.html.

People's Daily. (2009b). '3G licenses to be issued before the spring festival', *People's Daily*, January. [Online], Available: http://english.peopledaily.com.cn/90001/90776/90884/6567963.html.

People's Daily. (2011). 'China home to 1,200 foreign R&D Centers', *People's Daily Online*. 16 March. [Online], Available: http://english.peopledaily.com.cn/90001/90778/90861/6921243.html.

Petersmann, E.U. (2005). Reforming the World Trading System: Legitimacy, Efficacy and Democratic Governance. Oxford University Press.

Phillips, N. (2005). Globalising International Political Economy. Palgrave Macmillan.

Piggot, C. (2002). China in the World Economy: The Domestic Policy Challenges: Synbook Report. OCED.

Piller, F.T., Ihl, C. and Vossen, A. (2010). 'A typology of customer co-creation in the innovation process', *Social Science Research Network*, 4: 1–26.

Pisano, G.P. and Teece, D.J. (2007). 'How to capture value from innovation: shaping intellectual property and industry architecture', *California Management Review*, 50(1): 278–96.

Porter, M. (1990). The Competitive Advantage of Nations. Free Press.

Quan, Y., Haifeng, M. and Zhenhua, M. (2014). *China Focus: China's civil aviation industry growing with global integration.* English News CN.

Quarton, D. (2005). An International Design Standard for Offshore Wind Turbines: IEC 61400–3. Bristol, UK: Garrad Hassan and Partners, Ltd.

Quhong, S. (2011). Associate Professor in Innovation and Knowledge Management at Tsinghua University in Beijing. Personal Interview, Beijing, 15 April.

Qian, Z. (2012). 'Venture no further', *Global Times.* [Online], Available: http://www.globaltimes.cn/DesktopModules/DnnForge%20-%20NewsArticles/Print.aspx?tabid=99&tabmoduleid=94&articleId=743841&moduleId=405&PortalI D=0Radosevic, S. (1999). International Technology Transfer and Catch-Up in Economic Development. Edward Elgar.

Qing, L. Y. (2011).China's TD-LTE spreads across globe. ZD Net.

Rajaraman, J. (2011). 'China to Drive Development of 4G Network', *China Daily,* 20 May. [Online], Available: http://www.chinadaily.com.cn/business/2011–05/20/content_12546614.htm.

Rammel, C., Stagl, S. and Wilfing, H. (2007). 'Managing complex adaptive systems – A co-evolutionary perspective on natural resource management', *Ecological Economics,* (63): 9–21.

Ran, B. (2011). Beijing TV Station. Personal Interview. Beijing, 12 April.

Raven, R.P.J.M. (2005). Strategic Niche Management for Biomass. Eindhoven University of Technology.

Reddy, P. (2011). Global Innovation in Emerging Economies – Implications for Other Developing Countries. Institute for Economic Research, School of Economics and Management, Lund University, Sweden.

Reuters. (2011). China Mobile says Apple interested in TD-LTE, *Reuters,* March. [Online], Available: http://www.reuters.com/article/2011/03/04/china-npc-chinamobile-idUSHKU00046920110304.

Rivette, K. and Kline, D. (2000). 'Discovering new value in intellectual property', *Harvard Business Review,* January. [Online], Available: http://hbr.org/2000/01/discovering-new-value-in-intellectual-property/ar/1.

Robert, V. and Yoguel, G. (2010). 'Complex dynamics of economic development', Annual Conference for Development and Change.

Robertson, P.E. and Ye, L. (2013). On the Existence of a Middle Income Trap. Economics Discussion Paper 13.12.

Robinson, W.I. (2004). A Theory of Global Capitalism: Production, Class, and State in a Transnational World. Baltimore: Johns Hopkins University Press.

Robinson, M. (2008). 'Hybrid states: globalisation and the politics of state capacity', *Political Studies,* 56(3): 566–83.

Rohde and Schwarz. (2011). 'China Mobile Research Institute and Rohde & Schwarz to Conduct Joint Research on TD-LTE Testing Tech', Rohde & Schwarz, June. [Online], Available: http://www.antaranews.com/en/news/73315/china-mobile-research-institute-and-rohde – schwarz-to-conduct-joint-research-on-td-lte-testing-tech.

Roland Berger Strategy Consultants. (2003). From Middle Kingdom to Global Market Expansion Strategies and Success Factors for China's Emerging Multinationals. Roland Berger Strategy Consultants.

Roser, T., DeFillippi, R. and Samson, A. (2013) 'Managing your co-creation mix: co-creation ventures in distinctive contexts', *European Business Review*, 25(1): 20–41.

Ruiz, P.P. (2009). Procter and Gambles Connect and Develop. Open Innovate, Collaborative Networks and Engineering Design. [Online], Available: http://www.openinnovate.co.uk/pgs-connect-anddevelop/.

Rysavy Research. (2008). EDGE, HSPA, LTE: Broadband Innovation. Rysavy Research.

Rysavy Research. (2012). Mobile Broadband Explosion /4G Americas. Rysavy Research.

Saksena, A. (2009). High-Tech Industry: The Road to Profitability Through Global Integration and Collaboration. Cisco Internet Business Solutions Group (IBSG).

Salter, B. and Faulker, A. (2011). 'State strategies of governance in biomedical innovation: aligning conceptual approaches for understanding "Rising Powers" in the global context', *Global Health*, 7(3).

Sang, M., Lee, D., Olson, L. and Trimi, S. (2012). 'Co-innovation: convergenomics, collaboration, and co-creation for organisational values', *Management Decision*, 50(5): 817–31.

Santos, F. and Eisenhardt, K.M. (2006). *Constructing Markets and Organising Boundaries: Entrepreneurial Action in Nascent Fields*. Under Review Administrative Quarterly. Stanford University.

SASAC. (2010). Research and Markets: Report on China's Telecom Industry in 2010–2012. SASAC.

Scalise, G. (2004). Letter to Chairman on Subcommission on Information Technology and Competitiveness. Executive Office of the Presidents Council of Advisors. Washington.

Schilling, M.A. (2008). Strategic Management of Technological Innovation. McGraw-Hill Irwin.

Schuman, M. (2011). 'State capitalism vs the free market: which performs better?', Wall Street & Markets, *Time*. [Online], Available: http://business.time.com/2011/09/30/state-capitalism-vs-the-free-market-which-performs-better/.

Schwaag Serger, S. and Breidne, M. (2007). 'China's fifteen-year plan for science and technology: an assessment', *Asia Policy*, 4: 135–64.

Scissors, D. (2009). 'Deng undone the costs of halting market reform in China', *Foreign Affairs*, 88(3): 38–42.

Sesham, B. (2011). 'Essential LTE in focus as smartphone patent wars heat up', *International Business Times*. [Online], Available: http://www.ibtimes.com/essential-lte-focus-smartphone-patent-wars-heat-316278.

Sha, N. (2010a). 'China Mobile chairman Wang Jianzhou: innovation is more important than ever', C114, Online Blog, August. [Online], Available: http://www.cn-c114.net/576/a566551.html.

Sha, N. (2010b). 'MIIT: China Mobile TD-LTE added two equipment manufacturers, Ericsson and Motorola', C114, Online Blog, March. [Online], Available: http://www.cn-c114.net/576/a592453.html.

Sha, N. (2011a). 'STC launches the first 4G network-another breakthrough of ZTE and Huawei on TD-LTE', C114, Online Blog, October. [Online], Available: http://www.cn-c114.net/577/a650035.html.

Sha, N. (2011b). 'Huawei and ZTE gained the LTE commercial contracts', C114, Online Blog, December. [Online], Available: http://www.cn-c114.net/583/a571388.html.

Shapiro, C. (2000). 'Navigating the patent thicket: cross licenses, patent pools, and standard setting', *Innovation Policy and the Economy*, 1: 119–50.

Sharma, C. (2012). 'Global market update 2012: annual edition', *Always on Real-time Access Blog*. [Online], Available: http://www.chetansharma.com/blog/2012/04/30/global-mobile-marketupdate-2012-annual-edition/.

Shaughnessy, H. (2013). 'Tim Cook's new innovation strategy for Apple, right on time', *Forbes*. [Online], Available: http://www.forbes.com/sites/haydnshaughnessy/2013/07/02/tim-cooks-new-innovation-strategy-for-apple-right-on-time/.

Shen, D. and Shen, S. (2011). 'Volume production of TD-LTE handsets to not start until end-2012', *Digitimes*. [Online], Available: http://www.digitimes.com/news/a20110713PD226.html.

Shin, D.H., Choo, H. and Beom, K. (2011). 'Socio-technical dynamics in the development of next generation mobile network: translation beyond 3G', *Technological Forecasting & Social Change*, 78: 514–25.

Sigurdson, J. (2002). 'New Science and Technology & Innovation Developments in China', Conference, Proceedings of the Strata Consolidating Workshop Session 1: Globalisation-Strategies-of Multinational Corporations, International Research Co-operation and Implications for S & T Policies in Europe, Brussels, 22–23 April. [Online], Available: http://www.iso.org/iso/home/standards_development/resources-for-technical-work/stages_of_the_development_of_international_standards.htm.

Sit, V.F.S. and Lin, W.D. (2000). 'Restructuring and spatial change of China's auto industry under institutional reform and globalisation', *Globalisation Annuals of the Association of American Geographers*, 90(4): 653–73.

Sjoholm, F. and Lundin, N. (2013). 'Foreign firms and indigenous technology development in the People's Republic of China', *Asian Development Review*, 30(2): 49–75.

Soederberg, S.A. (2009). 'Critique of the diagnosis and cure for 'Enronitis': the Sarbanes-Oxley act and neoliberal governance of corporate America', *Critical Sociology*, 34(5): 657–80.

Spithoven, A. and Teirlinc, P. (2005). Beyond Borders: Internationalisation Of R&D And Policy Implications For Small Open Economics. Elsevier Science Publishing Company.

Staff Reporter. (2013). 'Ericsson hopes to improve performance in China through 4G market,' *Want China Times*, March. [Online], Available: http://www.wantchinatimes.com/news-subclass-cnt.aspx?id=20130301000003&cid=1206.

State Council. (2006). State Council of the People's Republic of China (SCPRC) Outline of the Long-Term National Plan for the Development of Science and Technology (2006–2020).

Steen, H.U. (2011). 'Indicators of development or dependency in disguise? Assessing domestic inventive capacity in South Korean and Chinese infrastructural ICT standards', *Telecommunications Policy*, 35: 663–80.

Steinbock, D. (2003). 'Globalisation of wireless value system: from geographic to strategic advantages', *Telecommunications Policy*, 27: 207–35.

Steinfeld, E. (2004). 'Chinese enterprise development and the challenge of global integration', in Yusuf, S., Altaf, M.A. and Nabeshima, K. (eds), Global Production

Networking and Technological Change in East Asia. The International Bank for Reconstruction and Development/The World Bank.

Stewart, J., Shen, X., Wang, C. and Graham, I. (2011). 'From 3G to 4G: standards and the development of mobile broadband in China', *Technology Analysis & Strategic Management*, 23(7): 773–88.

Strange, M. (2011). 'Discursivity of global governance vestiges of "democracy" in the World Trade Organisation', *Alternatives, Global, Local, Political*, 36(3): 240–56.

Strange, G. (2011). 'China's post-Listian rise: beyond radical globalisation theory and the political economy of neoliberal hegemony', *New Political Economy*, 16(5): 539–55.

Suchman, M.C. (1995). 'Managing legitimacy: strategic and institutional approaches', *Academy of Management Review*, 20(3): 571–611.

Sussmam. (2012). Complex Sociotechnical Systems: The Case for a New Field of Study. Charles L. Miller Lecture. MIT News.

Suttmeier, R. and Yao, X. (2004). China's Post-WTO Technology Policy: Standards, Software, and the Changing Nature of Techno-Nationalism. The National Bureau of Asian Research.

Suttmeier, R., Yao, X. and Tan, A. (2006). Standards of Power? Technology, Institutions, and Politics in the Development of China's National Standards Strategy. The National Bureau of Asian Research.

Swann, G.M.P. (2010a). The Economics of Standardisation: Report for the UK Department of Final Report for Standards and Technical Regulations Directorate. Manchester Business School. [Online], Available: https://www.gov.uk/government/uploads/system/uploads/attachment_data/file/16506/The_Economics_of_Standardization_-_in_English.pdf.

Swann, G.M.P. (2010b). 'International Standards and Trade: A Review of the Empirical Literature', OECD Trade Policy Working Papers, No. 97, OECD Publishing. doi: 10.1787/5kmdbg9xktwg-en.

Swissnex. (2009). A Quick Overview of the Science and Technology System in China. Swissnex, Shanghai.

Sun, L. (2011). 'Huawei tunes strategy to stay ahead in Europe', *China Daily*, July. [Online], Available: http://usa.chinadaily.com.cn/business/2011–07/22/content_12961215.htm.

Tabarrok, A. (2002). 'Patent theory versus patent law', *Policy*, 1(1): 1–24.

Tang, Li. and Yin, Li. (2011). Mini Country Report/P. R. China, Georgia Institute of Technology – ERAWATCH COUNTRY REPORTS 2011: China, [Online], Available: http://erawatch.jrc.ec.europa.eu/erawatch/export/sites/default/galleries/generic_files/file_0414.pdf.

Tang, L., and Shapira, P. 'Regional development and interregional collaboration in the growth of nanotechnology research in China.' *Scientometrics* 86(2): 299–315.

The Economist. (2010). 'Innovation in China patents, yes; ideas, maybe Chinese firms are filing lots of patents, how many represent good ideas?', *The Economist*, October.

Thorsten, B., Wolfgang, R. and Witte, M.J. (2002). Shaping Globalisation: The Role of Public Policy Networks. In Transparency: A Basis for Responsibility and Cooperation. Gütersloh: Bertelsmann Foundation Publishers.

Thun, E. (2006). Changing Lanes in China: Foreign Direct Investment and Auto Sector Development. Cambridge University Press.

Ting, L. and Shapira, P. (2011). 'Regional development and interregional collaboration in the growth of nanotechnology research in China'. *Scientometrics,* February, 86(2): 299–315.

Toner, P. (2011). Workforce Skills and Major Innovation: an Overview of Major Themes in the Literature. OECD Directorate for Science and technology (STI). Center for Educational Research and Innovation (CERI).

Toonen, W. (2013). The Globalised State: On the Impact of Globalisation on National State Capacity in the Perspective of Sub-Saharan Africa. BA Book.

Tsai, C.J. and Wang, J.H. (2011). 'How China institutional changes influence industry development? The case of TD-SCDMA industrialisation', Conference, Innovation, Strategy and Structure – Organisations, Institutions, Systems and Regions. Copenhagen Business School, Denmark, 15–17 June.

Tushman, M.L. and Nelson, R. (2012). 'Technology, organisations and innovation: an introduction', *Administrative Science Quarterly,* 35(1): 1–8.

Ukwandu, D.C. (2009). Water Use and Sustainable Development in South Africa. Master's Book. University of South Africa.

Unay, S. (2013). 'From engagement to contention: China in the global political economy,' *Perceptions,* 18(1): 129–53.

Underhill, G.R.D. and Zhang, X. (2005). 'The changing state-market condominium in East Asia: rethinking the political underpinnings of development,' *New Political Economy,* March, 10(1): 1–24.

United States Trade Commission, (2007). *China's Trade with the United States and the World.* CRS Report for Congress.

Updegrove, A. (2007). China Leads Developing Country Push for Balance in IP and Standards. Intellectual Property Watch.

Van Den Oord, A.J. (2010). The Ecology of Technology: the Co-Evolution of Technology and Organisation. PhD Book. Proefschrift.

van Someren, T.C. and van Someren-Wang, S. (2013). Innovative China Innovation Race Between East and West. Springer.

Vickery, G. and Wunschen-Vincent, S. (2009). 'R&D and Innovation in the ICT Sector: Toward Globalisation and Collaboration', The Global Information Technology Report 2008–2009. World Economic Forum.

Virki, T. (2010). 'Ericsson, Alcatel dominate LTE market: Dell'Oro,' *Reuters.* [Online], Available: http://www.reuters.com/article/2011/11/16/us-telecom-gear-research-idUSTRE7AF2RN20111116.

vom Haul, M. (2012). 'State capacity and inclusive development: new challenges and directions', ESID Working Paper, 02. [Online], Available: http://www.effective-states.org/_assets/documents/esid_wp_02_mvomhau.pdf.

Wade, R.H. (1990). Governing the Market. Princeton University Press.

Wade, R.H. (2003). 'Creating capitalisms', in Introduction to New Edition of Governing the Market. Princeton University Press.

Wai-chung Yeung, H. (2004). 'Strategic governance and economic diplomacy in China: the political economy of government-linked companies from Singapore,' *East Asia: An International Quarterly,* 21(1): 39–63.

Wai-chung Yeung, H. (2009). 'Regional development and the competitive dynamics of global production networks: an East Asian perspective', *Regional Studies,* 43(3): 325–51.

Wang, J.-H. (2006). 'China's dualist model on technological catching up: a comparative perspective', *Pacific Review,* 19(3): 385–403.

Wang, H. (2007). Policy Reforms and Foreign Direct Investment: The Case of the Chinese Automotive Industry. Ninth Gerpisa International Colloquium. Luxemburg.

Wang, H. (2008). 'Innovation in product architecture – A study of the Chinese automobile industry', *Asia Pacific Journal of Management*, 25: 509–35.

Wang, H. (2010). China's National Talent Plan: Key Measures and Objectives. Brookings.

Wang, J. (2011) cited in Hibberd, M., Middleton, J. and Weaver, P. (eds), *LTE Supplement*. Mobile Communications International.

Wang, H. (2011). Associate Professor, Chinese Academy of Social Science (CASS). Personal Interview, Shanghai, April.

Wang, Y. (2013). Will China Escape the Middle-income Trap? A Politico-economic Theory of Growth and State Capitalism. University of Zurich, (Job Market Paper).

Want China Times. (2013). 'Ericsson hopes to improve performance in China through 4G market', *Want China Times*. [Online], Available: http://www.wantchinatimes.com/news-subclass-cnt.aspx?id=20130301000003&cid=1206.

WBGU, (2011). *World in Transition – A Social Contract for Sustainability*. German Advisory Council on Global Change.

Weisberger, A. (2011a). 'Pyramid Research: Asia-Pacific to Be Global Leader in LTE by 2014 (following Wireless Intelligence Forecast)', *IEEE Communications Society*. [Online], Available: http://community.comsoc.org/blogs/ajwdct/pyramid-research-asia-pacific-be-global-leader-lte-2014-following-wireless-intelligence

Weisberger, A. (2011b). 'Huawei, ZTE and Ericsson to dominate telecom infrastructure equipment market – or not', *IEEE Communications Society*. [Online], Available: http://community.comsoc.org/blogs/ajwdct/huawei-zte-and-ericsson-dominate-telecom-infrastructure-equipment-market-or-no.

Weiss, L. (1998). The Myth of the Powerless State. Cornell University Press.

Weiss, L. (2005). 'Global governance, national strategies: how industrialised states make room to move under the WTO', *Review of International Political Economy*, 12(5): 723–49.

Wessner, C. W, and Wolff, A. W. (2012). Rising to the Challenge: U.S. Innovation Policy for Global Economy. Comparative National Innovation Policies: Best Practice for the 21st Century, Board on Science, Technology, and Economic Policy; Policy and Global Affairs; National Research Council.

Windsor, D. (2007). 'Toward a global theory of cross-border and multilevel corporate political activity', *Business and Society*, 46(2): 253–79.

WMGE, (2011). Flagship Report: World in Transition: a Social Contract for Sustainability. German Advisory for Climate Change.

Wolf, M. (1984). 'Two-edged sword: demands for developing counties', in Jagdish, N. and Ruggie, J.G. (eds), Power, Passions and Purpose: Prospects for North South Negotiations, 201–29, MIT Press.

Wolff, A. and Ballantine, D.B. (2006). 'China's drive toward innovation', in Wessner (ed.), Innovative Flanders: Innovation Policies for the 21st Century Report of a Symposium. The National Academies Press, Washington, DC.

Woods, B. (2011). 'ZTE grand memo hands-on: can it win the 'tablet' market?', *ZDNET*. [Online], Available: http://www.zdnet.com/zte-grand-memo-hands-on-can-it-win-the-phablet-market-7000011861/.

World Bank. (2013). *Building a Modern, Harmonious, and Creative Society*. Development Research Center of the State Council, the People's Republic of China.

Woo-Cummings, M. (1999). 'Introduction', in Woo-Cummings (ed.), The Developmental State, Cornell University Press.

Wortzel, L. M. (2013). *'The Dragon Extends its Reach: Chinese Military Power Goes Global'*. Potomac Books Incorporated.

Woyke, M. (2011). 'Identifying the tech leaders in LTE wireless patents', *Forbes*. [Online], Available: http://www.forbes.com/sites/elizabethwoyke/2011/09/21/identifying-the-tech-leaders-in-lte-wireless-patents/

Wu, J. and Callahan, J. (2005). 'Motive, form and function of international R&D alliances: evidence from the Chinese IT industry', *Journal of High Technology Management Research*, 16: 173–91.

Wu, D. and Zhao, F. (2007). 'Entry modes for international markets: case study of Huawei, a Chinese technology enterprise', *International Review of Business Research Papers*, 3(1): 183–96.

Xia, M. (2000). The Dual Developmental State: Development Strategy and Institutional Arrangements for China's Transition. Ashgate.

Xia, J. (2012). 'Competition and regulation in China's 3G/4G mobile communications industry – institutions, governance, and telecom SOEs', *Telecommunications Policy*, 36(7): 503–21.

Xiang, L. (2011). Vice President, Secretary General, 3G Industry Association. Personal Interview, 12 April.

Xiao, W. (2002). *Enterprise Leaders that Influence China's Economy*. Shenyang: Shenyang Press Ltd.

Xiaolan, F. and Xiong, H. (2011) 'Open innovation in China: policies and practices', *Journal of Science and Technology Policy in China*, 2(3): 196–218.

Xing, W. (2009). 'Huawei outshines rivals in telecom equipment mart', *People's Daily Online*, November. [Online], Available: http://www.chinadaily.com.cn/bizchina/2009–11/17/content_8983122.htm.

Xu, Y. (2003). The Economic Context of 3G Regional Overview China. Business briefing: Wireless Technology.

Yadong, L. (2007). 'From foreign investors to strategic insiders: shifting parameters, prescriptions and paradigms for MNCs in China', *Journal of World Business*, 42: 14–34.

Yan, H. (2007). 'The 3G standard setting strategy and indigenous innovation policy in China: Is T-SCDMA a flagship?', DRUID Working Paper no. 07–01. Danish Research Unit for Industrial Dynamics. [Online], Available: http://www.druid.dk/.

Yang, C.H. and Kuo, N.F. (2008). 'Trade-related influences, foreign intellectual property rights and outbound international patenting', *Research Policy*, 37: 446–59.

Yi, W. (2012). *China's Policy of Sustainable Development: Practices and Challenges*. Berlin Conference on Evidence for Sustainable Development: Evidence for Sustainable Development Semi-Plenary I: The Science-Policy Interface – Institutions and Boundary Organizations

Yoe, Y. (2009). 'Between owner and regulator: governing the business of China's telecommunications service industry', *China Quarterly*, 200: 1013–32.

Young, S. and Lan, P. (1997). 'Technology transfer to China through foreign direct investment', *Regional Studies*, 21(1): 669–79.

Young, D. and Yuabto, H. (2010). 'China's ZTE sees LTE business doubling or better in 2011', *Reuters*. [Online], Available: http://www.reuters.com/article/2010/11/17/zte-idUSHKU00035020101117.

Yu, J. (2004). Strategically Building Technological Capabilities in a Big Emerging Country. Institute of Policy and Management. Chinese Academy of Sciences.

Yu, J. (2011). 'From 3G to 4G: technology evolution and path dynamics in China's mobile telecommunication sector', *Technology Analysis & Strategic Management*, 23(10): 1079–93.

Yuan, Q., Mao, H. and Mao, Z. (2014). China focus: China's civil aviation industry growing with global integration. *English.news.cn. online*. [Online], Available: http://news.xinhuanet.com/english/china/2014–01/09/c_133031339.htm.

Yuen, L.C. (2006). New Trend of Chinese Mobile Communication Industry after China's WTO Accession. Hong Kong Baptist University Hong Kong.

Zadek, S. (2006). 'The logic of collaborative governance: corporate responsibility, accountability, and the social contract', Working Paper 17, Corporate Social Responsibility Initiative, Harvard Kennedy School, Cambridge.

Zang, B. (2002). 'Understanding China's telecommunications policymaking and reforms: a tale of transition toward liberalisation', *Telematics and Informatics*, 19: 331–49.

Zang, Y. (2009). Alliance-based Network View on Chinese Firms' Catching-up: Case Study of Huawei Technologies Co. Ltd. United Nations Working Paper.

Zang, C. (2011a). Doctor, Center for Informatisation Study, Chinese Academy of Social Sciences. Personal Interview, Beijing, 14 April.

Zang, M. (2011b). Associate Professor at School of Economics and Management (SEM), Tsinghua University. Personal Interview, Beijing, 15 April.

Zhang, H. K. (2008). 'What attracts Foreign Multinational Corporations to China?' *Contemporary Economic Policy*, 19(3): 336–346.

Zhang, J. and Liang, X.J. (2011). 'Business ecosystem strategies of mobile network operators in the 3G era: The case of China Mobile', *Telecommunications Policy*, 35: 56–171.

Zhang, X. and Prybutok, V. R. (2005). 'How the mobile communication markets differ in China, the U.S., and Europe.' Communications of the ACM – The Disappearing Computer. 48(3):111–114.

Zhao, M. (2010). 'Policy Complements to the Strengthening of IPRS in Developing Countries– China's Intellectual Property Environment: A Firm-Level Perspective', OECD Trade Policy Papers, No. 105, OECD Publishing. [Online], Available: http://dx.doi.org/10.1787/5km7fmtw4qmv-en.

Zhao, H. (2011). 'China's ZTE plans to be among top 3 LTE players by 2015', *iSuppi*. [Online], Available: http://www.isuppli.com/China-Electronics-Supply-Chain/MarketWatch/Pages/China's-ZTE-Plans-to-be-Among-Top-3-LTE-Players-by-2015.aspx.

Zheng, L. (2004). 'On the comparative advantage of Chinese industries', *Chinese Economy*, 37(2): 6–15.

Zheng, B. (2005). 'China's peaceful rise to great-power status', *Foreign Affairs*, 84(5): 18–24.

Zhen, L. and, Suny, Z. and Wrighty, B. (2012). *Patent subsidy and Patent Filing in China*. Penn State University.

Zhihong, W. (2006). 'Report: Carlyle likely to reduce Xugong stake', *China People's Daily*. [Online], Available: http://english.peopledaily.com.cn/200609/26/eng20060926_306486.htm.

Zhou, J. (2011). 'China to Drive Development of 4G Network', *China Daily*. 20 May.

Zhu, X. (2009). 'Science policy advisory in China: structures and social perspectives', in Ladikas, M. (ed.), Embedding Society in Science and Technology Policy: European and Chinese Perspectives. Office for Official Publications of the European Communities.

ZTE Corporation. (2010a). About ZTE: Company Overview. ZTE Corporation.

ZTE Corporation. (2010b). About ZTE: History. ZTE Corporation.

ZTE Corporation. (2010c). Milestones. ZTE Corporation.

ZTE Corporation. (2011). Company Aims to Achieve 10% Patent Share by 2012. ZTE Corporation.

ZTE, (2012). *ZTE Annual Company Report* 2012. ZTE.

Zutshi, A. (2009). 'Importance of global co-innovation networks: A TCS case study', IET Working Papers Series, Universidade Nova de Lisboa, IET-Research on Enterprise and Work Innovation, Faculty of Science and Technology.

Zweig, 2002:23).Internationalizing China: Domestic Interests and Global Linkages. Cornell University Press.

Index

GPSR Compliance
The European Union's (EU) General Product Safety Regulation (GPSR) is a set
of rules that requires consumer products to be safe and our obligations to
ensure this.

If you have any concerns about our products, you can contact us on

ProductSafety@springernature.com

In case Publisher is established outside the EU, the EU authorized
representative is:

Springer Nature Customer Service Center GmbH
Europaplatz 3
69115 Heidelberg, Germany